Use R!

Use R!

This series of inexpensive and focused books on R will publish shorter books aimed at practitioners. Books can discuss the use of R in a particular subject area (e.g., epidemiology, econometrics, psychometrics) or as it relates to statistical topics (e.g., missing data, longitudinal data). In most cases, books will combine LaTeX and R so that the code for figures and tables can be put on a website. Authors should assume a background as supplied by Dalgaard's Introductory Statistics with R or other introductory books so that each book does not repeat basic material.

More information about this series at http://www.springer.com/series/6991

Alfonso Zamora Saiz • Carlos Quesada González •
Lluís Hurtado Gil • Diego Mondéjar Ruiz

An Introduction to Data Analysis in R

Hands-on Coding, Data Mining, Visualization
and Statistics from Scratch

 Springer

Alfonso Zamora Saiz
Department of Mathematics Applied to ICT
Technical University of Madrid
Madrid, Spain

Carlos Quesada González
Department of Applied Mathematics
and Statistics
Universidad San Pablo CEU
Madrid, Spain

Lluís Hurtado Gil
eDreams ODIGEO
Barcelona, Spain

Diego Mondéjar Ruiz
Department of Applied Mathematics
and Statistics
Universidad San Pablo CEU
Madrid, Spain

ISSN 2197-5736 ISSN 2197-5744 (electronic)
Use R!
ISBN 978-3-030-48996-0 ISBN 978-3-030-48997-7 (eBook)
https://doi.org/10.1007/978-3-030-48997-7

Mathematics Subject Classification: 62-07, 68N15, 68N20, 68P05, 62-01, 68-01, 62-09, 62J05

This Springer imprint is published by the registered company Springer Nature Switzerland AG.
The registered company address is: Gewerbestrasse 11, 6330 Cham, Switzerland

Preface

One decade ago, the rise of big data was already a fact and several examples of successful data-driven companies were around. Nonetheless, its true importance and repercussion were still under question by many, especially on whether the new set of tools could be used not only by huge companies but also in a variety of aspects of life that could actually have an impact in society beyond increasing profits for businesses.

Of late, the fact that big data is useful beyond large-scale analysis is taken for granted and, in combination with other advances in mathematics, robotics, automation, and communication, data analysis has consolidated as the cornerstone of every new trend in business and society. Self-driving vehicles, facial recognition, and natural speech processing are some examples of data-driven technologies that are about to change the world (if not already have).

Even though this book does not dive into the latest methods of machine learning so as to understand how exactly a car is able to drive on its own, it is obvious that in this new data-oriented environment, the demand of graduates with proficient skills in the growing field of data analysis has increased notoriously. The need of STEM[1] professionals is clear, but students from other areas such as economy, business, social sciences, or law are equally important as the applications of data analysis affect all sectors of society. Typical STEM programs include courses syllabi with a strong core in mathematics, statistics, or even computing and programming, but it is frequently the case that other degree programs suffer from the lack of quantitative courses.

The archetypal job position in the next decade will require of the candidate to show proficiency in extracting conclusions from data through computing software, no matter the field or area of expertise. Of course, not all of them will be computer engineers or software developers in the strict sense, but they will be required to know the basics of coding to be able to develop data-oriented scripts and techniques

---

[1] Acronym for Science, Technology, Engineering, and Mathematics.

to analyze information. This implies the need to prepare alumni from other areas to handle data in their future jobs.

As professors, we have detected an increasing interest in data analysis from our colleagues coming from areas other than mathematics. Undergraduate students also express their desire to take extra credits in data-related subjects. Moreover, from our links to industry, we have been encouraged to instruct our students in analytical skills to face the era of data change. It is time to rethink undergraduate programs, especially in social sciences, and include courses addressing these kinds of competencies.

This text is conceived from the perspective of enabling students with no knowledge of data science to start their path in statistical computer-aided analysis with a broad scope. The book is the result on the one hand of our academic experience teaching in several undergraduate courses on business management, economics, business intelligence, and other master programs for the last years and, on the other hand, from our professional views as data scientists. It features the basics of programming and a series of mathematical and statistical tools that lead the reader from scratch to understanding the analysis of a database throughout the whole process of obtainment, preparation, study, and presentation of conclusions.

The programming language chosen for the book is **R**, a powerful yet simple way of handling data, specially designed for statistical analysis that eases out the first approach to coding, in contrast to other languages that might be equally useful but undoubtedly more difficult for an inexperienced reader.

The book contains sets of exercises for all sections and is thought to be taught in a semester course. It is self-contained, covering from the very basics in programming and even installing the required software to standard basic statistical methods to analyze data. All definitions in mathematics, statistics, and coding are provided; however, a reader with prior knowledge of statistics or already familiar with programming will find it much easier. In the same spirit, the usage of statistical models and coding tools is explained in detail but proofs or other non-practical details are not fully covered to its last extent. Ancillary materials are available through the data repository stored at https://github.com/DataAR/Data-Analysis-in-R.

Madrid, Spain   Alfonso Zamora Saiz
Madrid, Spain   Carlos Quesada González
Barcelona, Spain   Lluís Hurtado Gil
Madrid, Spain   Diego Mondéjar Ruiz
February 2020

Contents

List of Figures

Chapter 1
Introduction

Since the beginning of the twenty-first century, the humankind has witnessed the emergence of a new generation of mathematical and statistical tools that are reshaping the way of doing business and the future of society. Everything is data nowadays: company clients are tabulated pieces of data, laboratory experiments output is expressed as data, and our own history records through the internet are also made of data. And these data need to be treated, to be taken into account, to have all their important information extracted and to serve business, society, or ourselves. And that is the task of a *data analyst*.

Concepts such as data science, machine learning, artificial intelligence, and big data have invaded the conversations of entrepreneurs, students, big firms, and even families. It is often difficult to understand the difference between these words; as a matter of fact, they belong together inside the framework of modern data analysis and there is only a thin line dividing them. *Big data* is possibly the easiest to differentiate, although it refers to a complex scenario. This new paradigm is frequently explained through the "V's" model [10, 13], a term coined after several key words starting with *v*, that define the essence of big data. Technology solutions allow the storage of datasets of massive *volume*. Thanks to all sort of new sensors and sources of data retrieval, the *variety* of variables that are considered and their updating *velocity* is higher than anything that could have been imagined some decades ago. Such an enormous amount of diverse and instantaneous data has been proven of extreme *value* not only for commercial purposes but for farther more aspects of our everyday life. Such a high value encloses high risks and the use of sensitive data requires of a critical analysis on its *veracity* and ethical implications.

The implementation of the big data paradigm has been considerably fast and wide. Companies, institutions, and researchers showed their interest on properly implementing solutions for extensive amounts of data. According to Google Trends, searches of the term *big data* around the world increased by more than ten times

© Springer Nature Switzerland AG 2020
A. Zamora Saiz et al., *An Introduction to Data Analysis in R*, Use R!,
https://doi.org/10.1007/978-3-030-48997-7_1

between 2011 and 2015.[1] The development of techniques for data collection and management is still in progress, but its use is taken for granted. The processing of such amounts of data has required of advances on algorithms of statistical learning. This new generation of methodologies are called *machine learning*, the study of datasets with the aid of computers, not only to extract conclusions relying on statistical algorithms, but to set the computer with the ability of generating new conclusions with the arrival of new data.

By *data science* we refer to the wide range of procedures, methodologies, and applications related to the combined analysis of big data, understanding what data says, and machine learning, training devices being capable of learning as data arrives. Data science is not a classical scientific discipline, but a new paradigm through which we study reality, and can be (and has already been) applied with great success on multiple sciences, industries, or businesses. The conclusions obtained from these analyses and the decisions that are taken based on them are revolutionizing fields as different as astrophysics, finance, biotechnology, and industries ranging from automobile to pharmaceutical. The advances obtained from data analysis and its implementation on autonomous machines are producing a new generation of *artificial intelligence*, machines who are able to think and learn close to our human way, whose applications are yet to come in the shape of automation and robotics.

Let us step back to gain perspective of all these concepts and see why this is happening now and why this is as disruptive a revolution as everyone claims. Statistics started on the eighteenth century. It studies phenomena that cannot be described with deterministic functions, as physics do, and tries to approximate functional and non-functional relations for explanation or prediction. During the last centuries, procedures evolved and progressed and techniques were perfected, but this starting point of view remained unchanged: theoretically, relationships between variables are established, and empirical study confirms or rejects these models.

However, the invention and evolution of computers during the second half of the twentieth century yielded a significant change, which was not directly related to statistics at first. By early 2000, the cost for information storage and computing performance had decreased so much that completely new possibilities of interaction with computers were available, see Fig. 1.1[2] where it is worth noting that the scale is logarithmic.

In particular, computers were good enough not only to start processing big amounts of data, but also to test new methodologies. This opened a new paradigm which is twofold. Methods that could only be tested in small datasets, or that could have never been used some years ago because of the need of tremendous amounts of calculations, can be now performed in the blink of an eye on huge datasets with large amounts of variables. With the enhancement of calculations, the usage and applicability of classical methodologies have been improved and

---

[1] https://trends.google.es/trends/explore?date=2011-01-01%202015-03-03&q=\big%20data.
[2] Charts are created with data taken from [12].

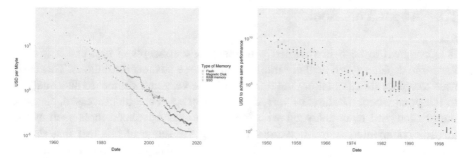

Fig. 1.1 Evolution of storage and computation costs

some ad hoc methods that require of computational muscle have been developed. But a more extreme situation has taken place; in order to guarantee that a certain algorithm actually describes an underlying reality of the world, it has been necessary to develop formal theories and make sure that assumptions are met, for which, validation techniques are used. Now it is possible to try certain numerical methods without going down to the exact probabilistic calculus and practically check whether an algorithm works or not. The most notorious example is neural networks, currently used in natural language processing, image recognition, or automation. If a neural network is set to distinguish photos of cats and dogs for example, and it does it correctly when facing pictures that the algorithm has never seen before (with a certain confidence level, say 95% of the times), then, undoubtedly, the algorithm is considered to be working by the scientific community. And there is no further discussion about the underlying model relating the variables, because it simply successes when explaining the phenomenon. This is not to say that the mathematical part should be disregarded; on the contrary, an extra effort should be done to catch up with reality, and many of the fundamental research in machine learning and artificial intelligence are currently on this path. But it is undeniable that effective new methods are springing because of this.

Not only computers are blazing fast and storage solutions are basically limitless also several ad hoc algorithms have appeared in the last years. The last factor is one of the last decade, the Internet of Things. IoT has enabled new possibilities due to cameras, GPS, phones, and plenty of other sensors that are constantly generating new data. This is one of the main sources that nurtures the big volume of data aforementioned. Finally, it is not to be forgotten that all these key features of data analysis are cheap, very cheap, or even free. All these aspects combined yield a new era in data analysis, where knowing statistics and computer programming is a must.

1.1   About This Book

It is common to be overwhelmed with many new concepts at first and it is even harder to depict oneself actually implementing these advanced solutions. This book tries to reduce this gap taking the reader from the very first contact to data analysis to the point where a dataset of various millions of rows can be processed, cleaned, understood, and even analyzed with the basic tools. Advanced tools such as the already commented neural networks are out of the scope of this text that should serve as an introductory step, leaving the reader precisely at the point of initiating the study of the advanced methods of data analysis.

There is plenty of quality literature on the topic, ranging from pure mathematical treatments to programming books. The particular approach of this book intends to be neither of them and somehow complements these previous texts. There are plenty of books covering the topic of data analysis with **R** from different points of view, some of the main references are gathered in the following list, that should not be taken as a fully comprehensive one. The references [14] and [1] do it from the pure programming perspective. Other approaches such as [3, 11], or [8] show the main statistical tools and algorithms for data science from various points of view and levels of difficulty. There are more mathematical texts such as [9] and [5] for the advanced reader and researcher going down to the heart of science.

Many of the previously mentioned texts focus just on the statistical part and tend to lack any comments on data obtainment or preprocessing, assuming that all data bases are already in a nice and convenient format. Other kind of books focus on the coding and take statistics for granted, assuming that the reader is already familiar with all algorithmic methods, like [4]. Very few, such as [7] cover a broader variety of aspects; basic knowledge of programming, data obtainment, preprocessing, and analysis; they are usually online versions and they frequently suffer from lack in elaboration due to explanations and variety of examples. Finally, some other very interesting books are focused on very specific topics such as visualization [2, 6].

One of the main goals of this book is to take the analyst from the very first contact with data to the final conclusion and presentation of results of analysis, and do that on a broad scope which does not assume a strong mathematical or statistical background, taking into account the variety of fields where data analysis occurs nowadays. We pay special attention to the different ways to obtain data and how to make it manageable before starting the analysis. The text includes a crash course on programming, plus the needed data processing tools placing the analyst at the starting point to extract information from data. Through statistical analysis plus visualization of this information, many conclusions can be extracted to understand the meaning of a database.

1.2   Overview

The book is structured as follows. The first chapter is devoted to a self-contained programming introduction to **R** with some history and the situation of **R** as a programming language in comparison to other options. Although this first contextual part can be seen as external to the data analyst we do believe that it will be profitable for a reader who never had any contact with programming languages and that has no notion about what the options are. This will help the reader into learning a fundamental tool such as a programming language. From there, switching to other languages can be achieved without remarkable additional effort. Once the introduction to programming is concluded the first notions of coding and data structures are explored in Sect. 2.2 and the chapter finishes in Sects. 2.3 and 2.4 with more specific features such as control structures and functions to automatize and standardize code processes.

Chapter 3 is devoted to everything related to datasets and it is rather innovative and new. It starts on Sect. 3.1 with a broad discussion on different aspects regarding databases to consider and keep in mind prior to any data analysis with special focus on data sources and file types. Section 3.2 is dedicated to the obtainment of data through different means such as APIs or scraping and Sect. 3.3 ends the chapter studying the details of preprocessing techniques, that is, the set of methods that are needed to convert raw data retrieved from different sources into conveniently formatted data that is ready to be analyzed. Throughout the chapter some examples are introduced and we will continue to work with them in the rest of the book.

Once the dataset is ready the book dives into exploratory analysis by means of visualization. Chapter 4 contains not only the different tools for plotting in **R** but also features rich discussions on the importance of visualizing, what is effective and what is not, and how a good plot can even enlighten what approach to use in the analysis to be carried out afterwards. It is divided into four sections, each one for each of the main plotting functions in **R**. A final section applies the knowledge to the previously referred datasets.

Finally, Chap. 5 illustrates the statistical concepts and how to use their methods to perform proficient analysis with the help of **R**. Section 5.1 starts off with the basics of descriptive statistics, then Sect. 5.2 goes further and explores inference, to check whether the data samples we use correspond to a known distribution or not, through the hypothesis test machinery. Then, Sect. 5.3 explores the multivariable relationships in a dataset with several features, finding correlations and fitting some basic linear models with explaining power about the case study. The chapter finishes with the analysis of the real life scenarios.

1.3   Data Analysis Examples

One of the major points of this book is to set the reader at the starting line to perform a real world analysis, which is precisely the reason why several real life data examples are included. Two of them are made of data which is initially scraped from two web pages and the other one is obtained from an API[3] and the three are then echoed during the rest of the text. The reproducibility of the first two is total, while the obtainment of the latter can be mimicked only with some additional effort. All these files can be accessed freely at our data repository https://github.com/DataAR/Data-Analysis-in-R.

Apart from that, **R** embeds several datasets ready for examples. They are widely used in many books or webpages and we use them as well since some of them are certainly well suited for education. However, remember that those sets are sometimes *too good*: it is extremely difficult to have datasets with such intuitive conclusions, clear patterns, and distinguishable relations. Additionally, nothing is said about the way they were created or what preprocessing was applied. Nevertheless, they constitute a good piece of information for the beginner analyst. The high value of the examples that we add follow the conception of the book; after finishing it, the reader should be able to conduct the analysis of real life data from the very beginning.

1.4   Notation and Terminology

Through the book we will refer to many technical concepts, both computer or mathematical and statistical related, which can be cumbersome at first.

R is a *programming language*, this is, a way to give instructions to the computer to perform a given task. These instructions are called a *code* which is presented in a color block in the text, while the *output* returned by the computer after realizing the task is shown below. The code can be inserted line by line in the R Studio *console* or run (line by line, or block by block) in a script file (see Sect. 2.1 for details). **R** is made of *libraries* comprehending different already implemented functions to complete an analysis job. Libraries have to be previously installed (once and forever) in **R** but only loaded during each **R** session to be used. A glossary of libraries and functions used throughout the book is also provided at the end.

Specially in Chap. 5, many statistical descriptions and mathematical formulae are included, trying to be, at the same time, self-contained but digestive for the novice reader. Whenever we have a *dataset* we understand that we have a *sample* extracted from a bigger *population* (e.g. to study European population we are given data from ten thousands of people from all countries). Each of these individuals comes

---

[3]Details on scraping and APIs are given in Sect. 3.2.

with a series of measures which are the *variables*, and which can be quantitative or qualitative. When taking about real data, we have particular values of these variables while when talking about the whole population, those variables are *random variables* that have a theoretical probability distribution. For example, if one of the features measured about the ten thousand people is height, we distinguish between a sample of 10,000 observations of height values from the theoretical random variable *height of all Europeans*, which can likely follow a normal distribution. The process of guessing properties about the population with the information provided with the sample is called *statistical inference*.

For multivariable statistics in Sect. 5.3, vectors are always considered column vectors. If X is a matrix, X^t denotes its transpose and X^{-1} its inverse, whenever it exists.

1.5 How to Download R and Rstudio

R can be downloaded for the various operative systems through the R Comprehensive Archive Network, CRAN, in the website https://cran.r-project.org. It is free software, continuously updated by the scientific community.

RStudio can be downloaded from https://www.rstudio.com, as an integrated environment very convenient to combine coding and visualizations of the output in the same window.

Upon completion of this book, last stable versions are R 3.6.2, called Dark and Stormy Night, released on December 12, 2019, and RStudio 1.2.5033, on December 3, 2019.

1.6 Repository

As mentioned above, all datasets used in this book can be easily accessed. While most of them are included in the different R packages used through this book, some additional files are provided for more realistic and complex analysis. These files may be large and can be freely downloaded at https://github.com/DataAR/Data-Analysis-in-R.

References

1. Adler, J. *R in a Nutshell*. O'Reilly Media, Inc., 2012.
2. BBC. BBC Visual and Data Journalism cookbook for R graphics. https://bbc.github.io/rcookbook/, 2018. [Online, accessed 2020-02-29].

3. B. S. Everitt and T. Hothorn. *A Handbook of Statistical Analyses Using R*. Chapman & Hall/CRC, 2006.
4. Hanck C., Arnold M., Gerber A. and Schmelzer M. Introduction to econometrics with r. https://www.econometrics-with-r.org/, 2018. [Online, accessed 2020-02-29].
5. Hastie, T., Tibshirani, R. and Friedman, J.H. *The elements of statistical learning: data mining, inference, and prediction*. Springer, Berlin, Germany, 2009.
6. K. Healy. *Data Visualization for Social Science: A practical introduction with R and ggplot2*. Princeton University Press, 2017.
7. Irizarry, R. Introduction to Data Science. Data Analysis and Prediction Algorithms with R. https://rafalab.github.io/dsbook/, 2019. [Online, accessed 2020-02-29].
8. James, G., Witten, D., Hastie, T. and Tibshirani, R. *An Introduction to Statistical Learning: With Applications in R*. Springer Publishing Company, Incorporated, New York, USA, 2014.
9. M. Kuhn and K. Johnson. *Applied predictive modeling*. Springer, New York, NY, 2013.
10. Lugmayr, A., Stockleben, B., Scheib, C., Mailaparampil, M., Mesia, N. and Ranta, H. A comprehensive survey on big-data research and its implications-what is really'new'in big data? It's cognitive big data! In *PACIS Proceedings*, page 248, 2016.
11. T. Mailund. *Beginning Data Science in R: Data Analysis, Visualization, and Modelling for the Data Scientist*. Apress, Berkeley, CA, USA, 1st edition, 2017.
12. McCallum, J. C. *Price-Performance of Computer Technology, chapter "Visualization" in the Computer Engineering Handbook, pp 4:1-18*. Vojin G. Oklobdzija, editor, CRC Press, Boca Raton, Florida, USA, 1st edition, 2002.
13. Patgiri, R. and Ahmed, A. Big data: The V's of the game changer paradigm. In *2016 IEEE 18th International Conference on High Performance Computing and Communications; IEEE 14th International Conference on Smart City; IEEE 2nd International Conference on Data Science and Systems (HPCC/SmartCity/DSS)*, pages 17–24, New Jersey, USA, 2016. IEEE.
14. Wickham, H. and Grolemund, G. *R for Data Science: Import, Tidy, Transform, Visualize, and Model Data*. O'Reilly Media, Inc., California, USA, 2017.

Chapter 2
Introduction to R

Learning a programming language is a difficult task and it requires a good effort of logical thinking to understand how a machine can interpret instructions to execute a step by step process. However, once the skill of programming in a specific language is acquired, learning yet another language becomes somewhat easier. For data analysts that are new to programming, **R** [19, 28] is not just another option among the wide universe of languages; it is a choice that pays off the learning investment very quickly, oriented to statistics in a very broad sense, that enables clear outputs of the programs designed.

The **R** programming language has become one of the most successful tools for data analysis. Since its development in the 1990s, it provides a comprehensive way to deal with numerical and textual data, implementing a variety of statistical tools to extract information from databases, focusing on the output, management, and exportation for further use. It is a rather easy tool (in comparison to other languages) that features all the elements involved in data analysis. Throughout the book, **R** will be used by means of the RStudio environment [31], a mainstream front-end software that eases out the implementation and coding.

This chapter is devoted to a first introduction to **R** starting off with a brief history of the language and then summarizing its characteristics. Afterwards we discuss the basics of programming with focus on the different data structures we can handle. Since **R** is a built-in array programming language it will be most friendly with vectors, matrices, and tables (called data frames), which will constitute the main objects to deal with. Control structures will set up the code in order to automatize tasks that vary upon the situation. Beyond the variety of functions implemented by default in **R** and tools to undertake different processes and calculations, it entitles analysts with the ability of constructing their own customized functions for the specific purposes of their projects.

This chapter acts as a *crash course* for programming in **R** that will set the reader at the starting point to use stronger tools and to manage and process data from different sources and nature, pursuing quantitative and qualitative conclusions.

© Springer Nature Switzerland AG 2020
A. Zamora Saiz et al., *An Introduction to Data Analysis in R*, Use R!,
https://doi.org/10.1007/978-3-030-48997-7_2

2.1   Programming Languages and R

Programming languages are the vehicles to encode computations and processes in machines and devices. Similar to a human language, a programming language comprehends a series of *syntactic, semantic,* and *grammatical* rules expressing ideas, concepts, and demands [13, 14, 32, 37]. It is a systematic way of structuring any procedure leading to a computer program: a set of instructions to be interpreted by a machine yielding the realization of a computing task. Semantics serve to give meaning to any word in the language and syntax establishes the form to connect the different words, while grammar determines how a character sentence is divided into words.

Despite the fact every single app or tool of computers, smartphones or other devices rely on programming, our purpose will be to use programming languages specifically to create algorithms. An *algorithm* is a finite number of sequenced instructions reproducing an operation that any person or machine can repeat, independently of their previous knowledge, as long as the rules and encoding that describe the algorithm are understood by that person, or by that machine [5, 21, 30]. Good examples are the division algorithm to compute the quotient and the remainder of the division between two numbers,[1] the Euclidean algorithm to extract the greatest common divisor or the Gauss method to solve linear systems of equations. All of them follow the same general pattern: for instance, in the division algorithm we start with two numbers, the dividend and the divisor (or numerator and denominator) and subtract repeatedly from the dividend multiples of the divisor until the remainder is smaller than the divisor. Once the method is correctly implemented, one just has to follow the rules properly and arrives to the correct solution. Even more, there is no way two different people will not get the same answer, the algorithm produces a univocal output.

To be able to reproduce these structures without external human support, society has developed during the last two centuries very sophisticated ways for machines and computers to understand these kind of instructions. These sets of commands that tell computers how to perform automated processes are precisely programming languages.

Sections 2.1.1 and 2.1.2 address topics that are not explicitly necessary for data analysis but serve as a great help for understanding the background of programming languages in a comprehensive and summarized way that might be of interest for all those who are completely new to coding. The eager reader with some record in programming can skip directly to Sects. 2.1.3 and 2.1.4 to start with the first steps.

---

[1]This algorithm is usually called long division method in US schools and many other places.

2.1.1 Some Remarks on the History of Programming Languages

In 1804, Joseph Marie Jacquard created a power loom named after him [25]. This was the first device to use coding to automatize a human task. It used punctured cards to manufacture textiles faster. The presence and absence of holes in the card encode sequences of processes to produce complex designs in fabric.

The *Analytical Engine* was designed in 1837 by Charles Babbage [4] to produce a mechanical computer storing 1000 numbers of 40 decimal digits with several functions (sum, subtraction, multiplication, division, square roots, and more). Mechanical calculators had been used since seventieth century [26], but this one was the first programmable calculator, where programs and data could be provided with punctured cards. Several improvements by Babbage in further years made possible to deal with polynomials and Gaussian elimination. Ada Lovelace, world-known as the first programmer in history, translated the memoir of Luigi Menabrea about Babbage's machine between 1842 and 1843 [24], specifying the mechanism to compute Bernoulli numbers.

Herman Hollerith [3] used the punctured cards to construct his *tabulating machine* which served to automatize and, hence, accelerate the process of finishing the 1890 US census [33]. He founded the Tabulating Machine Company which would be renamed as IBM in 1924 [27]. Years later, in 1936, a remarkable breakthrough in computation and programming was achieved with the Turing machine [10, 34]. This machine was a cornerstone in the history of computing, determining, in an abstract sense, which kind of computer machines can and cannot be constructed.

These mechanisms are examples of *assembly languages,*[2] low-level programming languages which depend essentially on the machine where they are used.[3] Since the 40s, there was an increasing desire to substitute assembly languages for the so-called *high-level languages*, able to be used in different machines. Konrad Zuse developed Plankalkül in 1942–1945 [22] and Corrado Böhm created in his PhD dissertation [6] the first language with a compiler, a computer program able to translate the source language into a language that the machine can understand. It would be in 1956 where the first programming language was made available to the public by IBM, the widely known and still used FORTRAN[4] [2]. Other languages back in the day were LISP [23] in 1958 and COBOL [9] in 1959, all of them still used nowadays.[5]

---

[2] Assembly languages are often abbreviated *asm.*

[3] Here, with machine we refer both to hardware, the architecture of the computer, and software, the operating system.

[4] FORTRAN is the acronym of FORmula TRANslation.

[5] Latest FORTRAN version was released on November 28, 2018, known as FORTRAN 2018, see https://wg5-fortran.org/f2018.html.

2.1.2   Paradigms and Structure

It is very difficult to attempt a general classification of programming languages. Generally, we distinguish between general-purpose programming languages (GSLs) for general use and domain-specific programming languages (DSLs) focused on one particular application. Examples of GLSs are FORTRAN, C or Python, while examples of DSLs are HTML for webpages or the mathematically oriented languages Matlab, Mathematica, or Maple. Besides that, languages are also classified by *programming paradigms* [35], depending on their characteristics. There is a myriad of paradigms, but the most common are the *imperative languages*, where the human being tells the machine what to do to change its state, and the *declarative languages*, where the human being tells the machine how the solution looks like instead of instructing how to achieve it. Among the imperative languages we find two subtypes, *procedural languages*, where the instructions are grouped into procedures, routines, or functions used as blocks of programming; and *object-oriented languages* where objects contain information (attributes) and interact between them with procedures (methods), instructions that combine both, the state of the objects and the interaction. Declarative languages can be divided into *functional*, where the resultant output is the value of implemented functions; *logic*, where the result is the evaluation of logic statements; and *mathematical*, where the answer comes from solving an optimization problem. Let us mention another paradigm, *array programming languages* (or vector or multidimensional languages) [20], from which **R** is a main exponent, realizing programming operations based on vectors, matrices, and higher-dimensional arrays. Matlab, Julia, or the NumPy extension to Python are other array programming languages. It is important to stress that all modern programming languages share many of these paradigms, not just one of them [15].

All programming languages rely on primitive structures or building blocks which are combined by syntax and semantic rules to form sentences which can be interpreted. The syntax takes care of which combination of symbols is correct and the semantic tells about the meaning of that combination. Once this is settled, the *type system* decides which variables, values, functions, structures are allowed in the language and classifies them into types [7]. Here we can distinguish between *static* (all types are defined once and for all, like C++ or C#) and *dynamic* languages (types can be modified along the time like Python, **R** or Julia). The possible interactions between the blocks are included in a *core library* containing all mechanisms of input and output plus the main functions and algorithms. Modern languages, as it is the case with **R**, have many additional libraries with procedures specifically designed for particular applications.

A programming language intends to imitate a natural language, with more or less complexity. However, the main difference is that common language is in constant evolution, and cannot be fully described nor established to last forever. Opposite to that idea, programming languages are finitely and fully specified as a set of rules from the very beginning, describing the future behavior of any correct sentence in the language; this comprehends syntax (as a formal grammar) and semantics (often written in natural language, as it happens with the C language), a description of the translator and a reference implementation from which all other codes will be derived.

The *implementation* of a programming language represents the way of writing code in another language that can be run in different machines. Implementation provides a way to write programs and execute them on one or more configurations of hardware and software. There are, broadly, two approaches to programming language implementation: *compilation* and *interpretation*. As we said before, a compiler translates a code in the programming language we are writing (the source code) into a lower level language understood by machines to create a program that can be executed to perform a real task. An interpreter is a software in the machine directly transforming and executing the code.

2.1.3 The R Language

To present **R** as a programming language it is necessary to mention its predecessor, S [8], an imperative, object-oriented and dynamic language devoted to statistics developed by John Chambers, Rick Becker, and Allan Wilks of Bell Laboratories, in 1976. Since the 80s, S-PLUS was sold as a commercial implementation by TIBCO Software.

R [16, 28] is a relatively new programming language, initially released in 1995 with its first beta version in 2000.[6] It has become widely used by data scientists, statisticians, and mathematicians to create statistical software and perform data analysis. **R** software is free and distributed as a package of the GNU[7] project[8] (a collection of computer software licensed under the GNU Project's own General Public License (GPL)). This language is available for the most common operative systems, Windows, Macintosh, Unix, and GNU/Linux. **R** was initially developed by Robert Gentlemand and Ross Ihaka in the University of Auckland in 1993 [19, 28] as a different implementation of S, combined with semantics strengths of another language, Scheme [12]. The name **R** for the language was coined after the initials of the forenames of the authors, plus the influence of the S language [17].

Since 2000, different versions have been made public.[9] Changes and improvements made in each version with respect to the last one, as well as updates and

---

[6]Message, by Peter Dalgaard, of the first beta version released https://stat.ethz.ch/pipermail/r-announce/2000/000127.html.

[7]GNU is a recursive acronym for "GNU's Not Unix."

[8]https://www.gnu.org.

[9]Upon completion of this book, last stable versions are R 3.6.2, called Dark and Stormy Night, released on December 12, 2019.

announcements are shown regularly at the *R project* website.[10] **R** is distributed as free software and can be downloaded from CRAN [16],[11] *The Comprehensive R Archive Network.*[12] There, a user can easily download and install more than 13,000 different packages developed by the worldwide community of users.[13] The amount and quality of these libraries should not be disregarded as it is in fact one of the strengths of this language. These packages cover an immense variety of statistical applications, methods, and algorithms. For an introductory to medium level text, such as this, we have to keep in mind that for every single of our challenges, there is probably already one or some commands that will solve it. Information and detailed documentation about **R** and libraries can be found via the *R project* website. In such a wide system of packages, functions, and resources, searching for help is essential. We will see how all these elements carry their own help file, with detailed descriptions of its content and examples of use. In addition, multiple forums can be found on the Internet, such as the **R**-help mailing list[14] or the *stackoverflow* webpage,[15] where thousands of users share questions and answers related with the user of **R**. Many of these users also contribute creating and developing new **R** packages around the world, the conference *useR!* being the annual official event for this purpose, held since 2004.[16] The **R** *Journal*[17] contains articles where innovations are described for developers and users.

As said, **R** is an imperative and object-oriented language. It also shares the functional and procedural paradigms, and it is an array programming language. Even though **R** is a programming language itself, for certain tasks with higher computational cost,[18] codes written in C, C++ or FORTRAN are called. As it is explained in the *R project* webpage, much alike the language S, **R** is an interpreted language developed in the real language of the machine, and that is why it does not need a compiler. Indeed, starting on Sect. 2.2 we will see how **R** is able to run pieces of code sequentially, without need of previous compilations. With this versatile features, users can easily create new procedures and functionalities, grouped together in libraries, to achieve new goals.

R has become one of the most used languages for data analysis because of its simplicity in comparison with other languages [11], the great variety of applications

---

[10]https://www.r-project.org.

[11]Visit https://stat.ethz.ch/pipermail/r-announce/1997/000001.html for the announcement by Kurt Hornik of the opening of CRAN site.

[12]https://cran.r-project.org/.

[13]In order to know the exact amount of available packages at a certain moment one can type nrow(available.packages()) on the **R** console..

[14]https://stat.ethz.ch/mailman/listinfo/r-help.

[15]https://stackoverflow.com/questions/tagged/r.

[16]https://www.r-project.org/conferences.html.

[17]https://journal.r-project.org.

[18]Computational costs are defined as the amount of time and memory needed to run an algorithm.

through its libraries,[19] the high quality of its graphical outputs and the detailed documentation. Surveys state that its popularity increased a lot during the past years, both in industry and in schools[20] and it is in a constant search for new horizons [18]. **R** has also influenced new streaming languages as it is the case of the programming language Julia,[21] intending to highly perform numerical analysis without compiling separately and, therefore, being faster.[22]

All this explains why **R** is today one of the most popular programming languages, with an increasing number of users and developers. Many of the most important technological companies and institutions use **R** for data analysis and modeling.[23] Learning **R** is one of the most demanded skills on any data driven job today.

2.1.4 RStudio

Learning a programming language can be hard for non-programmers because typing code in a black screen is tough. That is why modern programming has created *integrated environments* where coding is combined with graphical tools to quickly visualize data and graphics. The easiest way to use **R** is through the graphical integrated development environment *RStudio*, although other interfaces exist. RStudio is free and open source software and was founded by J.J. Allaire [1, 29]. RStudio can be used in two ways, RStudio desktop, working as an application in the computer and RStudio Server, using RStudio via remote access to a Linux server. Free distributions are available for Windows, macOS, and Linux. They can be downloaded from the team webpage.[24]

RStudio project started on 2010 and the first version available for users was released on 2011.[25] Besides creating a wonderful place to work with **R**, the RStudio team has contributed with many **R** packages. Among them, let us stress tidyverse, a system of packages specifically designed for data analysis, such as ggplot2. We can also mention RMarkdown, used to insert **R** code blocks into documents and knitr, to produce file reports by combining **R** code with LATEX and HTML. RStudio is partially written in the programming languages C++, Java, and JavaScript.

---

[19]Upon this book completion, the CRAN package repository features 15,368 available packages comprehending many possible extensions of the **R** core library.

[20]https://www.tiobe.com/tiobe-index/.

[21]https://julialang.org.

[22]https://julialang.org/blog/2012/02/why-we-created-julia © 2020 JuliaLang.org contributors.

[23]https://github.com/ThinkR-open/companies-using-r.

[24]https://www.rstudio.com.

[25]Announcement of RStudio release on February 28, 2011, https://blog.rstudio.com/2011/02/28/rstudio-new-open-source-ide-for-r/. Upon completion of this book, last stable version released is RStudio 1.2.5033, on December 3, 2019.

Fig. 2.1 RStudio interface. Provided by RStudio ® IDE open source

RStudio provides a great way to use the **R** programming language and interact with other software components. By default, when opening RStudio we find the screen divided into 4 parts (see Fig. 2.1). Top left is the *file* window where we write our code programs and are able to save it for further use. Bottom left is the *Console*, where instructions are executed, and the *Terminal*, to connect with the computer shell where commands in the computer language can be interpreted. Top right we can find three tabs: *Environment*, where all variables, datasets, and functions stored are shown; *History*, a summary of the actions taken in the session; and *Connections*, to connect with other data sources.[26] Bottom right we see 5 tabs: *Files*, to choose the working directory; *Plots*, to show the graphics generated; *Packages*, where libraries loaded are shown; *Help*, to get assistance with functions and commands; and *Viewer*, to have a look at data structures.

Instructions to be executed in RStudio can be introduced in two main ways. One is typing the corresponding instruction in the console, after the symbol >, and press enter. The result will be shown in the following line of the console, without >:

```
3 + 4
```

```
[1] 7
```

The symbol >, therefore, makes a difference between *input* and *output* in the code. Writing in the console has major drawbacks, code is not saved and different lines cannot be easily run at once. The recommended way to write code is using a file and run the whole or parts of it. There are two main *file types* to type code in RStudio. The basic script is the **R** *script* which produces a .R file. It looks like a blank

---

[26]For example, Oracle, OLBC, Spark, and many others.

Fig. 2.2 Window displaying an .R script. Provided by RStudio ® IDE open source

notebook (see Fig. 2.2) where code can be written line by line. We can choose to run the instructions line by line by clicking in the button Run[27] or selecting to execute a piece of code or the whole file.[28] When a code line is run, it is pasted in the console and performed, the output showing off again in the console, or in the Plot menu if it is a graphic, for example.

The extension .Rmd is the abbreviation of **R** *Markdown* (see Fig. 2.3), a file similar to .R with two main additional features. Firstly, code is inserted in blocks called *chunks* that are run block-separately and the output is displayed after each chunk, in a friendlier interface. This is ideal for testing and developing. Additionally, it allows to produce an HTML, PDF, or Microsoft Word report where text, code, and outputs are shown at once in a pretty and understandable format. To create the final document, compilation through the *knit button* is needed. The usage of markdowns requires additional libraries, see the following section for more information.

2.1.5   First Steps

Throughout this book, we will use the RStudio environment to write and execute programming codes and take advantage of all its features.

---

[27]A very useful shortcut is to use Control+Enter in PC, or Command+Enter in Mac, to run each code line.

[28]Have a look at the menu Code in RStudio for different run options.

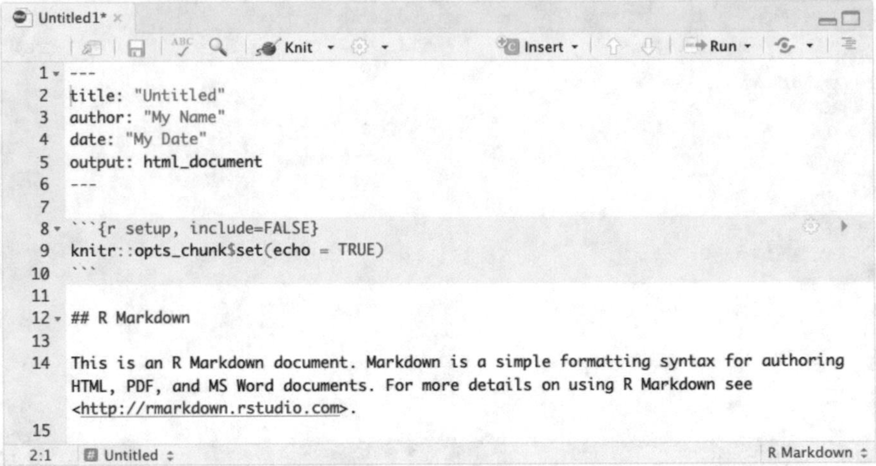

```
● Untitled1* ×                                                                        ─ □
       │ ▧ │ ⊟ │ ᴬᴮᶜ ᵠ │ ﹐Knit  ▾  ⚙  ▾            ﹒ Insert ▾  │ ⇧  ⇩ │ → Run ▾  │ ⟳ ▾ │ ≡
    1 ▾ ---
    2  title: "Untitled"
    3  author: "My Name"
    4  date: "My Date"
    5  output: html_document
    6  ---
    7
    8 ▾ ```{r setup, include=FALSE}                                              ⊚  ▸
    9  knitr::opts_chunk$set(echo = TRUE)
   10  ```
   11
   12 ▾ ## R Markdown
   13
   14  This is an R Markdown document. Markdown is a simple formatting syntax for authoring
        HTML, PDF, and MS Word documents. For more details on using R Markdown see
        <http://rmarkdown.rstudio.com>.
   15
   2:1      ▥ Untitled ⬍                                                    R Markdown ⬍
```

Fig. 2.3 Window displaying an .Rmd script. Provided by RStudio ® IDE open source

Input code will be presented in this fashion

```
for (i in 1 : 3) {
  print(2 * i)
}
```

and the produced output will be displayed exactly as it is shown in the console

```
[1] 2
[1] 4
[1] 6
```

The input code can be indistinctly copied and pasted to the console, .R and .Rmd files. It is left to the reader the choice of framework that best suits each situation. Sometimes the code will be too long for the width of the line and in these situations an explicit line break will be included, as RStudio does not notice this break and it makes the code suitable for copy-and-paste purposes.

```
temperatures <- c(18, 24, 26, 27, 19, 20, 22)
names(temperatures) <- c("Monday", "Tuesday", "Wednesday",
                         "Thursday", "Friday")
```

2.1.5.1   Libraries

As we said before, **R** is built as a corpus of *libraries* leading to different applications, numerical, graphical, intended for text processing, etc. All these packages (with the respective documentation) can be easily downloaded and installed from the CRAN repository. However, it is often the case that, while programming, an additional package is needed for our project. RStudio allows to download and install it directly from the console, by typing `install.packages()` and the name of the package[29] between quotation marks

```
install.packages("ggplot2")
```

It is very common that a particular package depends on previous packages to run correctly. If this is the case, then calling `install.packages()` for a particular package automatically detects the dependencies and downloads them as well (accompanied with an informative message in the output).

Once a package is downloaded, it is in our computer for future sessions but, in order to use the contents of a package, we need to call it in every single session with the command `library()` [30]

```
library(ggplot2)
```

Then, all data, functions, and components of that library are ready to be used. Once the package is installed we can see it in the Package tag (see Bottom right in Fig. 2.1). If the package is uploaded, we will see the tick activated. We can also load packages by ticking them in the Package Tag menu.

By typing

```
installed.packages()
```

we can see all packages currently installed in our computer and their characteristics: the package version, the **R** version needed for that package or dependencies upon other packages.

No automatic update tool is featured in **R** but the following code will check and update all packages, if possible:

```
update.packages(ask=FALSE)
```

For the rest of the book, many different packages will be used for data analysis. Whenever a new package is used, it will be stated by explicitly showing the load with the `library()` command, and it is left to the reader the installation of the package (with `install.packages()`), if necessary.

2.1.5.2 Writing Code

When writing code it is very important to be clear and follow a rational structure. Typically our code will be debugged, read, used, or modified by other people, therefore it must be very clear and follow standard and widely accepted rules. One of the recommended practices that can help to clarify the most are the *comments* in the code. These are sentences of natural language preceded by a hash symbol #. Comments will not be read when executing the code and serve to express additional ideas that help other humans understand what we programmed. As an example, have

---

[29]The package `ggplot2` will be called in Sect. 4.2 to create ellaborated plots in **R**.
[30]When calling `library()` quotation marks are not needed.

a look at this piece of code which prints a sentence and a number that are defined previously

```
# Script for printing a sentence and a number

sentence <- "The price is"  # defines the sentence
price <- 249.99  # defines the price
print(paste(sentence, price))  # prints both
```

```
[1] "The price is 249.99"
```

Even though there are several elements and commands that we do not understand yet, the explanation makes clear what is going on in every line.

As mentioned before, one of the best environments for code development is a **R** markdown file. To create one we only need to click on

```
File > New File > R Markdown...
```

In the opening window we will see different options but we will only mention the first one, Document. After choosing a name for our project and writing the author name we can select the output format of our work: HTML, PDF, or Word. Then we click OK. A new file is open in RStudio and it contains a few help comments. We recommend visiting the help links for greater details and utilities on markdowns.[31]

R markdowns use mainly two packages: rmarkdown and knitr. The first one is already installed on RStudio, but we need to install the second one. **R** markdowns allow us to mix formatted text with code chunks and pictures. Whenever we create a new **R** markdown file, a default help code is given. Delete it and write instead

```
Script for printing a sentence and a number

```{r}
sentence <- "The price is"  # defines the sentence
price <- 249.99  # defines the price
print(paste(sentence, price))  # prints both
```
```

The first sentence, outside the code chunk, is read as a simple formatted text. Code section will be started by ```{r} and ended by ```. Everything found between these marks will be interpreted as a code and therefore, comments need the hash symbol. Then we can save our file as normally. After installing knitr the Knit button is enabled at the top of the file window. If we click on it, the file is compiled into an HTML, PDF, or Word, depending on our initial choice. A new file is produced and automatically open, where we can see *name of the file* as main title, our text, and code in boxes. Notice how the code output is produced in a separated box preceded by double ##.

The **R** markdown also provides further utilities that we will not explore in this work. Just mention the green arrow at the top right of the code chunk. We can click

---

[31] https://rmarkdown.rstudio.com.

on it to run the current chunk and the code output will appear after calculation below it. That makes markdowns a feasible option to mix your code with long comments and text structure.

2.1.5.3 Searching for Help

When learning a programming language, the pile of words, commands, and concepts to learn grows exponentially and even expert programmers do not memorize all functions of every language. Small changes from one language to another make it really difficult to keep everything in mind and false friends show up, the same as with natural language. For these situations **R** and the Internet provide different help utilities that can be highly useful for our work.

The first help tool is the **R** command help(), from the built-in package utils. By typing help and the word we want to know about between parenthesis (or just the question mark followed by the word to look up), the **R** documentation is shown in the Help tag (see the bottom right area in Fig. 2.1)

```
help(matrix)
?matrix
```

In this case, we are told that the text matrix is a function from the base package (one of the core libraries installed with **R** by default) and a description of how to use it, with its syntax, arguments, examples, and bibliography. This command can be used as well with datasets and packages. Try for example

```
help(package="ggplot2")
```

to read the help files of package ggplot2 (explained in Sect. 4.2) or

```
help(diamonds, package="ggplot2")
```

to learn about the diamonds dataset included in ggplot2 package.

All **R** packages include detailed documentation files and we strongly recommend visiting them as a complement for this book and for further details. However, these files might be considered too arid for inexperienced users. Their content is usually limited to brief technical descriptions and definitions of **R** objects. When we need to learn about a new package and how to make use of it in our analysis, we may need more pedagogical resources. This book is meant to be self-explanatory for a wide range of commands, packages, and other facilities, but the possibilities that **R** offers us are much bigger. Fortunately, many packages include vignettes and user manuals that fulfill this function. Type vignette() to see on RStudio a list of vignettes from all of your installed packages. With browseVignettes() this list is opened in our browser, which facilitates their lecture. Another similar help tool are demos, code demonstrations that show examples and graphs created by a command or package. Type demo() for a list. For example, type

```
demo(graphics)
```

to see different example plots created with the graphics package (explained in Sect. 4.1).

These help tools to require the name of a known command, dataset, or package to show related information. Nevertheless, **R** users constantly face problems for which efficient solutions have already been created and published. The authors of this book strongly recommend the reader to *use published R functions* rather than creating his own. Available packages can be trusted as optimal solutions for the engaged problems, more efficient than any function written by an inexperienced programmer. In addition, using public functions standardizes our code, making it easier to read other users' scripts. Searching for published solutions, installing and using them save time when we know where to look for solutions. The problem is how to find these published solutions for our problems.

Two help commands can be used to search for information in our computer. Command apropos() returns a list of all objects available in our session containing a character string. If we have previously estimated a linear regression using command lm() (see Sect. 5.3.2 for more information), and we preserve the objects related with this calculation, we can list them with apropos("lm"). If we want instead to search among the documentation of the installed packages, we will use help.search() or ?? followed by the string of interest. Type help.search("lm") to see a list of all pages citing command lm. However, these engines search only inside our **R** current session, while we may need new solutions contained in packages unknown to us. For an external search we can use RSiteSearch() on character strings. Now, typing RSiteSearch("histogram") we will open in our browser a detailed list of more than 6000 vignettes and other documents using histograms in **R**. This is an especially interesting tool when we know the name of a technique but we do not know how to make use of it on **R**. In most cases, we will find these techniques already implemented in **R**.

These functions provide official support for basically all commands, but help goes way beyond that. The **R** community has developed the *CRAN Task Views* and encyclopedic list of packages classified by topics.[32] If we are interested in a scientific topic, such as clustering, machine learning, or econometrics, we can find these categories and click for an extensive list of packages with brief definitions. A little search on this web page might help us to find the most suitable package for our problem. This list is usually updated to the most recent and advanced contributions in **R**. A similar help source is the Frequent Asked Questions on **R** [17], a list of the most common questions asked by **R** users, and related to its general use.

If all these options fail, we can always make the questions ourselves in forums or, more commonly, search for users who faced the same problems in the past. In the **R** mailing list[33] and Internet forums such as *stackoverflow*[34] we will find

---

[32]We can access this list at https://cran.r-project.org/web/views/.

[33]https://www.r-project.org/mail.html.

[34]https://stackoverflow.com/.

thousands of solved questions that can be of great use for us as well. With a simple search on Google adding "R" or "CRAN" we can find published on the Internet the conversations held by users that solved these problems in the past. If our question is rather specific and we do not find a solution, we can always log in the **R** mailing list and send an e-mail or open a new thread in a public forum. From the authors experience, we can receive a relevant answer in less than 48 h. The messages have to follow certain rules for reproducibility, including a detailed explanation of our problem and a sample of our code.[35]

2.2 Data Structures

As data analysts, the basic ingredient for any study is the data we account for. Data can encode various types of information and characteristics of very different nature about the case study. It can be presented as numbers, which are, in turn, integer or decimal. It can also be made out of words, a single word, a whole sentence, or even a chain of characters, like a code or a password. Another frequent format is a feature of the type yes/no, a dichotomous information. Once every single piece of information is understood, there might be an underlying structure in data, a sequence of values, a table of cross-data for two features, or a list of ordered categories for certain individuals.

All programming languages are designed to handle different kinds of information. All of them work with the basic types of data mentioned above, but then, depending on the purpose of the language, more complex objects arise. Some structures, as vectors or matrices, are inherent to most modern languages. Others, such as factors or data frames, are very useful contributions of **R** for data analysis. In this section we will go over most of the *data structures* that are implemented in **R**, ranging from variables storing a single number or character to data frames with information of different types. Understanding such structures and being proficient in their usage is the very first step into data analysis.

2.2.1 Variables and Basic Arithmetic

First, we learn how to store basic information and perform elementary calculations and operations with it. This will lead to the mathematical concept of variable, which can be thought of as a box with a given name storing a certain piece of data.

---

[35]Learn how to make your questions reading https://stackoverflow.com/questions/5963269/how-to-make-a-great-r-reproducible-example and https://www.r-project.org/posting-guide.html.

The very first thing we will do is use **R** as a normal *calculator* to evaluate simple expressions. For example, if we type in the console the number

```
3
```

and we press enter, we obtain this as a result,

```
[1] 3
```

meaning that the evaluation of the expression we introduced is three, and the output has just one line, indicated between brackets. Similarly, we can type basic operations such as *sum* and *subtraction*, and get the corresponding results

```
3 + 4
```

```
[1] 7
```

```
3 - 5
```

```
[1] -2
```

To introduce *products* and *quotients*, we use the symbols ∗ and /, respectively,

```
2 * 6
```

```
[1] 12
```

```
5 / 3
```

```
[1] 1.666667
```

and the *module* operation %% returns the reminder of the division and %/% returns the integer division,

```
15 %% 2
```

```
[1] 1
```

```
15 %/% 2
```

```
[1] 7
```

The next step is to define *variables* whose purpose is to store values to be retrieved for later use. It is worth mentioning that they are stored in the *RAM memory*,[36] a type of memory that is extremely fast to read or write, but that cannot be preserved once the computer is switched off. Nowadays, most computers have enough amount of RAM memory to undertake most of the calculations in this book, therefore we do not care much about it when dealing with basic calculations and small databases. However, for tables of several hundreds of millions of observations, variable storage becomes a major problem that will be discussed in depth in Sect. 3.3.

---

[36]The acronym RAM stands for Random-access memory.

The way to assign a value to a variable is by writing the name of the variable, the symbol < -, and the value we want to store

```
x <- 4
```

Symbol < - is the assign symbol,[37] used to store any **R** object into a new one. Then, if we type the name of the variable, we recover the value stored

```
x
```

```
[1]  4
```

We can use previous variables to perform calculations. For example, define a new variable

```
y <- 5
```

and use it together with the previous one x to get the sum of both,

```
z <- x + y
z
```

```
[1]  9
```

The three basic data types commented in the introduction are implemented in **R**: numeric, character, and logical. In order to check what kind of variable is being used, we use the command class(). Class *numeric* corresponds to numbers

```
number.decimal <- 4.3
class(number.decimal)
```

```
[1]  "numeric"
```

For very large or very small numbers, **R** can use scientific notation to save space without loss of accuracy. For example, number 27,300 can be written as well as $2.73 \cdot 10^4$, or in **R**

```
2.73e4
```

```
[1]  27300
```

where e separates the decimal and the exponent. For very small numbers we can use as well this notation

```
1 / 2.73e4
```

```
[1]  3.663004e-05
```

meaning $3.663004 \cdot 10^{-5}$ or 0.00003663004.

---

[37]The same result is achieved by typing assign("x",4).

A variable of class *character* stores character chains. To introduce characters we have to use quotation marks

```
word <- "Hello"
class(word)
```

```
[1] "character"
```

The *logical* class encodes Boolean[38] values TRUE and FALSE

```
yes.no <- TRUE
class(yes.no)
```

```
[1] "logical"
```

Logical values are useful to perform arithmetics. Value TRUE also corresponds to value 1 and FALSE to value 0,[39] and therefore, we can perform operations with them

```
TRUE + TRUE
```

```
[1] 2
```

```
TRUE + FALSE
```

```
[1] 1
```

```
TRUE * FALSE
```

```
[1] 0
```

```
FALSE * FALSE
```

```
[1] 0
```

It is worth mentioning two *special* data types. When assigning NULL to a variable we convert it into an empty object. This can be counter intuitive but it is very common in programming and represents an entity in the workspace containing no information at all

```
null.object <- NULL
null.object
```

```
NULL
```

---

[38] A *Boolean* expression is a data type whose possible values are either TRUE or FALSE. It is named after the mathematician George Bool.

[39] For simplicity, logical values can also be written as T or F. We will use the full word or the initial letter indistinctively throughout the book.

The other type is the NA, meaning a *non-available* entry and usually representing a missing value in data analysis. An NA can be thought as a gap instead of a particular value for a variable

```
missing <- NA
missing
```

```
[1]  NA
```

Note that, opposite to NULL, NA *stores information* (in spite of being no information at all!) and this is why we see one row in the variable missing but no rows in null.object. Basic arithmetic with them results on NA's again:[40]

```
1 + NA
```

```
[1]  NA
```

```
TRUE - NA
```

```
[1]  NA
```

2.2.2 Vectors

A *vector* is an array of elements. We can create vectors in **R** with the command c() and the elements inside separated by commas

```
vectors.numbers <- c(1, 2, 3, 4)
vectors.numbers
```

```
[1]  1 2 3 4
```

The length of the vector we just created can be retrieved by

```
length(vector.numbers)
```

```
[1]  4
```

We can also create vectors of character or logical class elements

```
vector.characters <- c("Today", "Yesterday", "Tomorrow")
vector.characters
```

```
[1] "Today"         "Yesterday"          "Tomorrow"
```

---

[40]In Sect. 5.1 we will see how to remove these NAs when performing calculations over vectors containing them, with the argument na.rm.

```
vector.logicals <- c(TRUE, FALSE)
vector.logicals
```

```
[1]  TRUE FALSE
```

The command c() comes from *combination* [36] and it is used to combine elements that must be of homogeneous class. When the type of data is not the same, the information is *coerced* into the same class. This means that **R** tries to create a vector of the least general type among the possibilities, following the hierarchy

$$\text{logical} < \text{numeric} < \text{character}.$$

For example, c(3.4, "Goodbye") results in c("3.4", "Goodbye") where the 3.4 of numeric type has been coerced into the character type. Or c(TRUE, 3.4) will be stored as c(1, 3.4). Missing NAs can be also coerced to other types with c()

```
c(NA, 3, 4)
```

```
[1] NA   3   4
```

Numerical vectors are of special interest for **R** users. Many times these vectors can be defined by a formula or sequence and we just want to reproduce them. For the simplest cases, several ways are provided to create them. With the colon command and notation a:b, we can create a vector with all consecutive numbers between a and b, both included.

```
4 : 12
```

```
[1]  4  5  6  7  8  9 10 11 12
```

If, instead, we want to repeat a number or group of numbers, we will use command rep(). The default use is to repeat the first argument the number of times indicated by the second argument

```
rep(c(3, -1, 0.5), times=3)
```

```
[1]  3.0 -1.0  0.5  3.0 -1.0  0.5  3.0 -1.0  0.5
```

or we can instead repeat each element in the first argument a given number of times.

```
rep(c(3, -1, 0.5), each=3)
```

```
[1]  3.0  3.0  3.0 -1.0 -1.0 -1.0  0.5  0.5  0.5
```

A combination of both uses can be done by assigning a vector to argument times with as many elements as the first one. The resulting vector will repeat each element as many times as given by times

```
rep(c(3, -1, 0.5), times=c(2, 1, 3))
```

```
[1]  3.0  3.0 -1.0  0.5  0.5  0.5
```

And finally, we can ask **R** to repeat the elements of a vector as many times as necessary to reach a given length

```
rep(c(3, -1, 0.5), length.out=8)
```

```
[1]  3.0 -1.0  0.5  3.0 -1.0  0.5  3.0 -1.0
```

Another, more elaborated, way of generating vectors is through function `seq()`. As we will see in Sect. 4.1.2 with the histogram plots, this is a useful command to separate an interval into same length subintervals. We only need to specify the beginning of the interval, the end, and the increment size

```
seq(0, 10, 2.5)
```

```
[1]  0.0  2.5  5.0  7.5 10.0
```

Once vectors are created, the coordinates can be labeled with names. This will be useful when referring to a particular position of the vector. For example, create a vector with temperatures from Monday to Friday and assign labels to the day names by using the command `names()`. The variable `names(temperatures)` contains a character vector with the five day names.

```
temperatures <- c(28, 29, 27, 27, 30)
names(temperatures) <- c("Monday", "Tuesday", "Wednesday",
                         "Thursday", "Friday")
temperatures
```

```
   Monday   Tuesday Wednesday  Thursday    Friday
       28        29        27        27        30
```

Note that there is no longer a `[1]` in the output but the structured vector with the labels. Now, create another vector with the rains for those days, and take advantage of temperatures names to label rains days

```
rains <- c(0, 5, 6, 0, 2)
names(rains) <- names(temperatures)
```

or, equivalently, create an auxiliary vector of days, and label our vectors with it

```
days <- c("Monday", "Tuesday", "Wednesday", "Thursday", "Friday")
names(rains) <- days
rains
```

```
   Monday   Tuesday Wednesday  Thursday    Friday
        0         5         6         0         2
```

Given two vectors, we can perform different *operations* in each coordinate, in the same way we sum vectors in geometry. The sum of a vector with a number

returns the sum of each coordinate by that number. For example, add 273.15 to
`temperatures` (in Celsius degrees) to get the values in Kelvin degrees

```
temperatures + 273.15
```

```
   Monday    Tuesday Wednesday  Thursday     Friday
   301.15     302.15    300.15    300.15     303.15
```

where note that the answer keeps labels from both summands. Combining both sum
and multiplication, we can calculate the values of temperatures, now in Fahrenheit
degrees[41]

```
temperatures * 1.8 + 32
```

```
   Monday    Tuesday Wednesday  Thursday     Friday
    82.4       84.2      80.6      80.6       86.0
```

Similarly, subtraction, division, and powers can be achieved by typing the characters
`-`, `/`, and `^`. For example, this expression attempts to calculate the ratio between
rainfall and temperature, expressed as liters per square meter to Kelvin degrees

```
rains / (temperatures + 273.15)
```

```
     Monday      Tuesday   Wednesday     Thursday       Friday
0.000000000 0.016548072 0.019990005 0.000000000 0.006597394
```

Observe how division is performed coordinate-by-coordinate, and vector labels are
preserved in the answer.

The function `sum()` returns the sum of all coordinates in a vector,

```
total.rains <- sum(rains)
total.rains
```

```
[1] 13
```

There are different ways to access and *select* the elements in a vector. The basic
one is to type between brackets the position of the element

```
rains[2]
```

```
Tuesday
      5
```

To select more than one position, we can use vector notation

```
rains.beginning.week <- rains[c(1, 2)]
rains.beginning.week
```

```
 Monday Tuesday
      0       5
```

---

[41]The conversion between Celsius degrees °C and Fahrenheit degrees $°F$ is $°F = 1.8 \times °C + 32$.
To go from Celsius to Kelvin we just shift the zero in the scale to 273.15.

Referring to consecutive positions can be done through the colon command `:`. Then we use this vector as positions

```
temperature.midweek <- temperatures[2 : 4]
temperature.midweek
```

```
  Tuesday Wednesday  Thursday
       29        27        27
```

Once we have labeled vectors, the names can be used when selecting as follows

```
temperature.thursday <- temperatures["Thursday"]
temperature.thursday
```

```
Thursday
      27
```

```
rains.beginning.week <- rains[c("Monday", "Tuesday")]
rains.beginning.week
```

```
 Monday Tuesday
      0       5
```

Another essential relation between variables that are ordered is *comparing* them. The following are the different ways to compare two values in **R**:

- < for less than
- > is greater than
- <= is less than or equal to
- >= is greater than or equal to
- == is equal to each other
- != is not equal to each other

If we want to check whether a certain relationship between two values is true or false, we type the corresponding logical expression whose evaluation will be a logical value

```
6 > 3
```

```
[1] TRUE
```

```
4 == 5
```

```
[1] FALSE
```

When comparing a vector with a value, or with another vector, we obtain a vector of logicals, each of them checking the logical condition for each element

```
rains > 0
```

```
   Monday   Tuesday Wednesday  Thursday    Friday
    FALSE      TRUE      TRUE     FALSE      TRUE
```

We can as well consider multiple conditions. The symbol & means *and*, i.e. both conditions should be satisfied to get TRUE overall, while the symbol | means *or*, i.e. at least one should be satisfied to get TRUE overall:

```
3 == 4 & 3 == 3
```

```
[1] FALSE
```

```
3 == 4 | 3 == 3
```

```
[1] TRUE
```

```
3 < 5 & 4 > 2
```

```
[1] TRUE
```

```
TRUE & TRUE & FALSE
```

```
[1] FALSE
```

The function xor(condition1, condition2) returns TRUE if one, and just one of the conditions are TRUE; it returns FALSE if both are FALSE or both are TRUE

```
xor(3 > 2, FALSE)
```

```
[1] TRUE
```

```
xor(TRUE, 4 == 4)
```

```
[1] FALSE
```

These comparatives are useful when selecting those elements of a vector verifying certain property. For example, selecting those days with no rain,

```
not.rainy.days <- rains == 0
not.rainy.days
```

```
   Monday   Tuesday Wednesday  Thursday    Friday
     TRUE     FALSE     FALSE      TRUE     FALSE
```

or selecting the rains for those days with temperature above or equal to 29°,

```
hot.days <- temperatures >= 29
hot.days
```

```
   Monday   Tuesday Wednesday  Thursday    Friday
    FALSE      TRUE     FALSE     FALSE      TRUE
```

```
rains[hot.days]
```

```
Tuesday  Friday
      5       2
```

When performing selection in a vector based on the verification of a particular condition, there are three singularly useful functions. One is `all()`, checking whether the conditions are satisfied by all elements of the vector. For example, with the following code we can tell that there is, at least, 1 day whose temperature is lower than 28°

```
all(temperatures >= 28)
```

```
[1] FALSE
```

If, instead, we want to check if the condition is satisfied by, at least, one element of the vector, we can use the command `any()`. Now, we see that there are days with temperature equaling 30°

```
any(temperatures == 30)
```

```
[1] TRUE
```

The command `which()`, however, finds the actual positions of those elements satisfying the condition and returns the TRUE ones as a vector. In the example, we find that the third and fourth day reached 27°

```
which(temperatures == 27)
```

```
Wednesday   Thursday
3           4
```

A fluid use of indices and positions in vectors is an instrumental skill for an efficient use of more advanced datasets. In the following sections we will see how to sort a dataset, but first we need to learn how to sort vectors with command `order()`. It returns the positions of the sorted elements in a vector, from minimum to maximum. With that, we can reorder the vector as `vector[positions of sorted elements]` as in this example:

```
some.vector <- c(3, 7, 9, 6, 2, 8)
order(some.vector)
```

```
[1] 5 1 4 2 6 3
```

This means that the smallest value is in the fifth position, the second smallest in the first, and so on. Then, to retrieve the ordered vector just introduce

```
some.vector[order(some.vector)]
```

```
[1] 2 3 6 7 8 9
```

The same output can be obtained in a much simpler fashion with the code `sort(some.vector)` although this approach does not allow to keep track of the positions in the original vector, which can be very valuable.

In addition to this, we can calculate basic operations with vectors, such as the sum of the whole vector or obtaining the maximum and minimum elements

```
sum(temperatures)
```

```
[1] 141
```

```
max(temperatures)
```

```
[1] 30
```

```
min(temperatures)
```

```
[1] 27
```

or, both extremes with function `range()`, which returns a vector

```
range(temperatures)
```

```
[1] 27 30
```

With a combination of the previous commands, we can obtain the day of the maximum temperature

```
which(temperatures == max(temperatures))
```

```
Friday
    5
```

Note that these commands are inherently endowed with a parenthesis that is filled with an input. All commands that feature different outcomes for different inputs are known as *functions*. In the context of programming languages, inputs for functions are given by arguments in the same way functions in mathematics depend on variables. For example, the mathematical function

$$f(x, y) = 2x^2 - y + 1$$

is a function depending on two variables, x and y. Each time we assign particular values to x and y, we obtain an output value of $f(x, y)$ as a return, for example, when $x = 2$, $y = 5$, f takes the value $f(2, 5) = 4$. Functions in **R** will follow the same structure, choices on the values of the arguments will modify the output. We will see more on functions in Sect. 2.4.

2.2.3 Matrices

After learning how to use vectors to store a series of data of certain length, we step into *matrices*. Matrices are one of the most important tools in mathematics, used to encode a 2-dimensional array of numbers, each of them named by their row and column coordinates. In programming, matrices play the same role, with the advantage of being used to store tables of data of the same kind, not necessarily numbers.

The command `matrix()` allows to create matrices and it is the first function we use that has more than one argument. The syntax of this function is

```
matrix(data=NA, nrow=1, ncol=1, byrow=FALSE, dimnames=NULL)
```

where the different arguments are:

- `data` is the vector of data we want to include in the matrix
- `nrow` is the number of rows
- `ncol` is the number of columns
- `byrow` indicates whether we fill the matrix with the data vector by rows or columns; by default it is filled by columns
- `dimnames` allows to include a list (see Sect. 2.2.6) of 2 elements containing names for rows and columns; there are no names by default.

Note that all arguments in the function `matrix()` have a value by default. This is a general feature of **R** or any programming language, unless a different value is specified for those arguments, they take the default value.

We now create our first matrix; a 3×3 matrix of the consecutive numbers from 1 to 9, disposed by rows. To do that, we need to type the command `matrix()` with the needed arguments filled up

```
matrix(1 : 9, byrow=TRUE, nrow=3)
```

```
     [,1] [,2] [,3]
[1,]    1    2    3
[2,]    4    5    6
[3,]    7    8    9
```

Once the vector containing the data to fill the matrix is provided together with the number of rows or columns, the length of the other dimension is automatically deduced by the **R** language. If the vector length is not divisible by the number of rows or columns that we indicate, we will receive a warning message, but still **R** will fill out the matrix with the vector, starting over from the beginning if necessary,

```
matrix(1 : 15, nrow=2)
```

```
data length [15] is not a sub-multiple or multiple of the number
ofrows [2]
     [,1] [,2] [,3] [,4] [,5] [,6] [,7] [,8]
[1,]    1    3    5    7    9   11   13   15
[2,]    2    4    6    8   10   12   14    1
```

We can create matrices of character data. For example, this matrix of the months in the year

```
matrix(c("January", "February", "March", "April", "May", "June",
        "July", "August", "September", "October", "November",
        "December"), nrow=3, byrow=TRUE)
```

```
     [,1]          [,2]         [,3]        [,4]
[1,] "January"     "February"   "March"     "April"
[2,] "May"         "June"       "July"      "August"
[3,] "September"   "October"    "November"  "December"
```

Also, matrices that are made of logical values,

```
matrix(c(TRUE, TRUE, FALSE, TRUE), ncol=2)
```

```
     [,1]  [,2]
[1,] TRUE  FALSE
[2,] TRUE  TRUE
```

Once we have a matrix, we can add particular names to each row and column with the commands `rownames()` and `colnames()`,

```
climate <- matrix(c(temperatures, rains), byrow=TRUE, nrow=2)
rownames(climate) <- c("Temperatures", "Rains")
colnames(climate) <- days
climate
```

	Monday	Tuesday	Wednesday	Thursday	Friday
Temperatures	28	29	27	27	30
Rains	0	5	6	0	2

The dimensions, number of rows and columns, of a matrix can be obtained by the function `dim()`

```
dim(climate)
```

```
[1] 2 5
```

To add new data as a row or column to a given matrix we use `rbind()` and `cbind()`, respectively. Let us add a new row with wind speed data to the `climate` matrix

```
Winds <- c(30, 25, 22, 24, 18)
rbind(climate, Winds)
```

	Monday	Tuesday	Wednesday	Thursday	Friday
Temperatures	28	29	27	27	30
Rains	0	5	6	0	2
Winds	30	25	22	24	18

Using the function rowSums(), which sums all values in each row of a matrix and returns the vector of sums, we can add a column of total values to the matrix

```
total.climate <- rbind(climate, Winds)
totals <- rowSums(climate)
cbind(climate, totals)
```

	Monday	Tuesday	Wednesday	Thursday	Friday	totals
Temperatures	28	29	27	27	30	141
Rains	0	5	6	0	2	13

Observe that, when adding the vector totals to the matrix climate, the new column takes the value of the vector.

Selection of matrix elements is similar to vectors. Instead of one coordinate, now we have two of them, the first referring to the row and the second referring to the column. For example, to get the value in row 2 and column 3

```
total.climate[2, 3]
```

```
[1] 6
```

To get all values in column 4 we leave in blank the row position

```
total.climate[, 4]
```

Temperatures	Rains	Winds
27	0	24

We proceed similarly with rows to extract the temperatures in row 1

```
total.climate[1, ]
```

Monday	Tuesday	Wednesday	Thursday	Friday
28	29	27	27	30

We can also refer to rows and columns by their names

```
total.climate[, "Friday"]
```

Temperatures	Rains	Winds
30	2	18

```
total.climate["Rains", ]
```

Monday	Tuesday	Wednesday	Thursday	Friday
0	5	6	0	2

Using the function `mean()`, which calculates the arithmetic mean of a vector,[42] we can get the average temperature by getting the first matrix row

```
ave.temperature <- mean(total.climate[1, ])
ave.temperature
```

```
[1] 28.2
```

Matrices allow to perform the same arithmetic calculations than we did with vectors, element by element. The following operation multiplies each item in the matrix by 3

```
climate * 3
```

	Monday	Tuesday	Wednesday	Thursday	Friday
Temperatures	84	87	81	81	90
Rains	0	15	18	0	6

This squares each element in the matrix

```
climate ^ 2
```

	Monday	Tuesday	Wednesday	Thursday	Friday
Temperatures	784	841	729	729	900
Rains	0	25	36	0	4

To calculate the ratio between rains and temperatures, we can divide one row by another

```
climate[2, ] / climate[1, ]
```

```
    Monday     Tuesday   Wednesday    Thursday      Friday
0.00000000  0.17241379  0.22222222  0.00000000  0.06666667
```

2.2.4 Factors

Qualitative or *categorical* variables are those variables whose values are not numerical, but categories. In **R**, we encode this type of information by means of *factors*. Given a vector representing the values taken by a categorical variable and applying the command `factor()` to the vector, it is converted into a vector of factors.

Create a vector named `sizes` that contains nine observations from a variable taking three different values, `Small`, `Medium`, and `Big`. Now, each element contains the corresponding word of the given observation

```
sizes <- c("Small", "Big", "Big", "Medium", "Medium", "Small",
           "Medium", "Small", "Small")
sizes
```

---

[42]See Sect. 5.1.1 for an explanation of the arithmetic mean and other statistical measures.

```
[1]  "Small"  "Big"     "Big"     "Medium" "Medium" "Small"
[7]  "Medium" "Small"   "Small"
```

Typing `summary()` [43] of the vector we see that the only information we get is the number of elements and their class

```
summary(sizes)
```

```
   Length     Class      Mode
        9 character character
```

However, if we convert the vector `sizes` to a factor vector by using `factor()`, the resulting structure recognizes how many different values the variable `size` takes, and how many times each value appears, as it can be seen with `summary()`

```
factor.sizes <- factor(sizes)
factor.sizes
```

```
[1] Small  Big    Big    Medium Medium Small  Medium Small  Small
Levels: Big Medium Small
```

```
summary(factor.sizes)
```

```
   Big Medium  Small
     2      3      4
```

The *categories* of a factor vector are now displayed along with the vector and the command `summary()` shows the number of entries of each factor. We can obtain these values with the command `levels()`

```
levels(factor.sizes)
```

```
[1] "Big"    "Medium" "Small"
```

Qualitative variables can be measured in a *nominal* or an *ordinal* scale. For example, colors follow a nominal scale, because there is no hierarchy between the values and they do not admit any ordering:[44]

```
my.colors <- c("Orange", "Red", "Red", "Yellow", "Blue", "Green")
factor.colors <- factor(my.colors)
factor.colors
```

```
[1] Orange Red    Red    Yellow Blue   Green
Levels: Blue Green Orange Red Yellow
```

---

[43] `summary()` is one of the most robust and powerful commands in **R**. Almost all kind of structures can be passed as an argument of this command and it will usually provide plenty of information.

[44] Everything can be ordered, alphabetically for example, but nominal scales have no meaningful order related to anything intrinsic to the nature of the variable.

However, some other qualitative variables admit an ordering, in other words they are measured in an ordinal scale. When that is the case, the argument `ordered` is set to the value TRUE and the order of levels is specified with a vector in the argument `levels`, so that the variable is factored in an ordinal scale:

```
sizes2 <- c("Small", "Big", "Big", "Medium", "Medium", "Small",
            "Medium", "Small", "Small")
factor.sizes2 <- factor(sizes, ordered=TRUE,
                        levels=c("Small", "Medium", "Big"))
factor.sizes2
```

```
[1] Small  Big    Big    Medium Medium Small  Medium Small  Small
Levels: Small < Medium < Big
```

Factors are useful when collecting ordered information from surveys and different names are assigned to the factors. Imagine that a survey collects information on the gender of a sample as "M" for male and "F" for female. By factoring the vector, it is then easy to gather the frequencies of the qualitative variable

```
survey.vector <- c("M", "F", "F", "M", "M", "F", "M", "M")
factor.survey.vector <- factor(survey.vector)
factor.survey.vector
```

```
[1] M F F M M F M M
Levels: F M
```

```
summary(factor.survey.vector)
```

```
F M
3 5
```

These outputs are sorted out into an increasing ordering of frequencies, therefore F appears before M because 3 is smaller than 5. We can overwrite the whole vector by just changing the `levels` to the real words of the qualitative values of gender, using the previous order

```
levels(factor.survey.vector) <- c("Female", "Male")
factor.survey.vector
```

```
[1] Male    Female Female Male    Male    Female Male    Male
Levels: Female Male
```

Factoring variables in an ordinal scale allows to compare the categories themselves and the elements of the factor vector. When the variable is unordered there is no way to compare different values

```
factor.survey.vector[1] < factor.survey.vector[2]
```

```
[1] NA
Warning message:
In Ops.factor(factor.survey.vector[1], factor.survey.vector[2]) :
< not meaningful for factors[1] NA
```

However, in the case of ordered variables, where a particular ordering is specified, we can compare values

```
small <- factor.sizes2[1]
medium <- factor.sizes2[4]
big <- factor.sizes2[2]
small < medium
```

```
[1] TRUE
```

```
small < big
```

```
[1] TRUE
```

```
medium > big
```

```
[1] FALSE
```

```
factor.sizes2[3] < factor.sizes2[5]
```

```
[1] FALSE
```

2.2.5 Data Frames

So far, we have seen vectors that encode series of data and matrices of 2-dimensional data. These structures work with data of the same type: numbers, characters, and logical values.[45]

When performing data analysis we will be forced to deal with data of different types at the same time. For example, a medical study of 300 patients where, for each individual, fever is measured as a non-integer number, prescribed medication is stored as an unordered factor, hospital admission is represented by date and time, location is a character, and the response to a disease test is a true or false logic variable. *Data frames* are the main data structure in **R**, and will be used to store data tables where, unlike matrices, data might belong to different classes. Once created, accessing different variables and particular elements in the table will be much easier.

There are some datasets already stored as data frames in **R** for educational purposes, to be used as examples, and they can be found in the datasets package. To get information on this package we can type

```
?datasets
library(help="datasets")
```

---

[45]Thanks to the combination command c(), if data are of different types, all of them are stored in the most general type admitting all kinds appearing in the structure.

which displays the documentation of the package technicalities and contents. This
dataset is preloaded by default when **R** is opened so, in order to choose a particular
example, we just type its name. In addition to `datasets`, some packages also carry
their own datasets and we can use them after installing the package. We can see all
available datasets in our computer with

```
data()
```

We choose a particular dataset as an example and have a look at its structure. `OrchardSprays` collects data from an experiment to assess the potency
components of orchard sprays in repelling honeybees. We can access to this
information by using the help `?OrchardSprays`, or obtaining the whole table
with `OrchardSprays`

```
?OrchardSprays
OrchardSprays
```

This data frame has four variables: `rowpos` and `colpos` encoding the row and
column of the design, `treatment` is a factored qualitative variable encoding the
treatment and taking eight values from "A" to "H", and `decrease` which is the
numerical response to the treatment. We can have a look at the first six rows of the
data frame with `head()`

```
head(OrchardSprays)
```

```
  decrease rowpos colpos treatment
1       57      1      1         D
2       95      2      1         E
3        8      3      1         B
4       69      4      1         H
5       92      5      1         G
6       90      6      1         F
```

or get information about its structure with `str()`

```
str(OrchardSprays)
```

```
'data.frame':   64 obs. of  4 variables:
$ decrease : num  57 95 8 69 92 90 15 2 84 6 ...
$ rowpos   : num  1 2 3 4 5 6 7 8 1 2 ...
$ colpos   : num  1 1 1 1 1 1 1 1 2 2 ...
$ treatment: Factor w/ 8 levels "A","B","C","D",..: 4 5 2 8 7 ...
```

New data frames can be created with the function `data.frame()` that uses,
as arguments, the vectors that will form the columns of the data frame.[46] In the
example, we see a data frame collecting some information about the authors of this
book[47]

---

[46]Unlike matrices, if the column lengths to be included in the data frame are not the same, the
function returns an error and a data frame filling the gaps is not created.

[47]In Spain and other countries, two family names are used, preserving both the last name of the
father and the mother.

```
name <- c("Alfonso", "Carlos", "Lluis", "Diego")
last.name <- c("Zamora", "Quesada", "Hurtado", "Mondejar")
second.last.name <- c("Saiz", "Gonzalez", "Gil", "Ruiz")
age <- c(33, 32, 30, 37)
phd <- c("math", "math", "physics", "math")
office <- c(4, 14, 6, 8)
from.madrid <- c(FALSE, TRUE, FALSE, TRUE)

professors <- data.frame(name, last.name, second.last.name, age,
                         phd, office, from.madrid)
str(professors)
```

```
'data.frame':    4 obs. of  7 variables:
 $ name            : Factor w/ 4 levels "Alfonso",..: 1 2 4
 $ last.name       : Factor w/ 4 levels "Hurtado",..: 4 3
 $ second.last.name: Factor w/ 4 levels "Gil",..: 4 2 1 3
 $ age             : num  33 32 30 37
 $ phd             : Factor w/ 2 levels "math","physics": 1 1 2 1
 $ office          : num  4 14 6 8
 $ from.madrid     : logi  FALSE TRUE FALSE TRUE
```

We can *select* elements of a data frame the same way we did with matrices, accessing them by their row and column coordinates,

```
professors[2, 3]
```

```
[1] Gonzalez
Levels: Gil Gonzalez Ruiz Saiz
```

```
professors[1, ]
```

```
     name last.name second.last.name age  phd office from.madrid
1 Alfonso    Zamora             Saiz  33 math      4       FALSE
```

```
professors[, 2]
```

```
[1] Zamora   Quesada   Hurtado   Mondejar
Levels: Hurtado Mondejar Quesada Zamora
```

```
professors[1 : 2, ]
```

```
     name last.name second.last.name age  phd office from.madrid
1 Alfonso    Zamora             Saiz  33 math      4       FALSE
2  Carlos   Quesada         Gonzalez  32 math     14        TRUE
```

One of the best advantages of data frames is that columns have names coming from the variables of the table. We can use these names to access different elements by

using the $ symbol after the data frame name, and then selecting a particular element
of the column vector

```
professors$second.last.name
```

```
[1] Saiz       Gonzalez Gil        Ruiz
Levels: Gil Gonzalez Ruiz Saiz
```

```
professors$office[3]
```

```
[1]  6
```

We can use the logical content of some data frame variables to subset it

```
madrileans <- professors$from.madrid
professors[madrileans, ]
```

```
      name last.name second.last.name age  phd office from.madrid
2 Carlos   Quesada           Gonzalez  32 math     14        TRUE
4  Diego   Mondejar              Ruiz  37 math      8        TRUE
```

or use the command subset (data_frame, subset=logical_condition)

```
subset(professors, subset=age < 31)
```

```
      name last.name second.last.name age     phd office from.madrid
3 Lluis    Hurtado                Gil  30 physics      6       FALSE
```

```
subset(professors, subset=phd == "math")
```

```
       name last.name second.last.name age  phd office from.madrid
1 Alfonso     Zamora             Saiz   33 math      4       FALSE
2  Carlos    Quesada         Gonzalez   32 math     14        TRUE
4   Diego    Mondejar             Ruiz  37 math      8        TRUE
```

A basic utility is to sort data frames with respect to a certain variable, as long as
that variable has any ordering. This can be done with function order() as seen in
Sect. 2.2.2

```
positions <- order(professors$age)
professors[positions, ]
```

```
       name last.name second.last.name age      phd office
3   Lluis    Hurtado               Gil   30 physics      6
2  Carlos    Quesada          Gonzalez   32    math     14
1 Alfonso     Zamora             Saiz    33    math      4
4   Diego    Mondejar             Ruiz   37    math      8
from.madrid
      FALSE
       TRUE
      FALSE
       TRUE
```

Note that, even though the rows have been reordered, the original numbering of the rows is preserved and displayed in the left most column of the output, so that **R** can keep track of the original information.

It is often the case that we have data stored in a different structure, for example as a matrix, and our desire is to convert it into a data.frame to enjoy the advantages of it. This can be done with as.data.frame(). Suppose we have a matrix containing the first 9 consecutive numbers,

```
mydata.as.matrix <- matrix(1 : 9, nrow=3)
mydata.as.matrix
```

```
     [,1] [,2] [,3]
[1,]    1    4    7
[2,]    2    5    8
[3,]    3    6    9
```

and we want to have that piece of data in the form of a data.frame. We can convert it with

```
mydata.as.data.frame <- as.data.frame(mydata.as.matrix)
mydata.as.data.frame
```

```
  V1 V2 V3
1  1  4  7
2  2  5  8
3  3  6  9
```

Once we have the matrix as a data.frame we can access their columns as if it were the variables, with the $ symbol and the corresponding name[48]

```
mydata.as.data.frame$V1
```

```
[1] 1 2 3
```

The topic of data frames is a really rich one. It is in fact one of the main topics to focus on in modern data analysis, and plenty of other interactions can be performed. One of the reasons why **R** is specially appreciated, is its ease for dealing with large datasets. Learning **R** allows us to rapidly access our information and operate with it.[49] Data frames will be further studied in connection with databases in Sect. 3.3.

2.2.6   Lists

Sometimes, we need to perform analysis over very heterogeneous pieces of information. Some of them can be stored as vectors, some admit matrices, some

---

[48]When applying as.data.frame, unless otherwise specified, the default names of the variables are $V1$, $V2$, etc., meaning variable 1, variable 2, etc.

[49]Some **R** packages are specially designed for dealing with datasets, such as tibble and data.table, we will explore the later one in Chap. 3.

can be processed as data frames, etc. However, it is important to have a structure giving room to all of them together because, for example, they refer to the same individual to study. As an example, consider that we are comparing traffic in 10 cities and, for each of them, we have basic data (single variables as name, GPS coordinates, country, population), information on the climatology (stored as a matrix of temperatures, rains, pressure, and daylight hours), and a data frame of main traffic spots with several characteristics (location, tourism attractions nearby, traffic flux, or restrictions to contaminant vehicles). For each city we have the same information, but each singular piece is of different nature, hence a new structure allowing this comes in handy, *lists*.

Lists can be conceived as *vectors* made of elements which can be of different nature: different lengths, sizes, types, etc. The way to create a list is with the command list() and passing as arguments all pieces we want to store in the list. We now create a list made out of the vector days, the factor vector factor.sizes, and the data frame climate

```
new.list <- list(days, factor.sizes, climate)
new.list
```

```
[[1]]
[1] "Monday"    "Tuesday"    "Wednesday" "Thursday"   "Friday"

[[2]]
[1] Small  Big   Big    Medium Medium Small  Medium Small  Small
Levels: Big Medium Small

[[3]]
             Monday Tuesday Wednesday Thursday Friday
Temperatures     28      29        27       27     30
Rains             0       5         6        0      2
```

In a list, we can see whole elements as if it were a vector

```
new.list[1]
```

```
[[1]]
[1] "Monday"    "Tuesday"    "Wednesday" "Thursday"   "Friday"
```

and, more concretely, we can point elements by first referring to the position in the list using double brackets and then following with the particular structure inherited from the data that is being handled

```
new.list[[1]][2]
```

```
[1] "Tuesday"
```

```
new.list[[3]][1, 3 : 4]
```

```
Wednesday  Thursday
       27        27
```

We can also give names directly to the elements in the list

```
new.list <- list(the.days=days, the.factors=factor.sizes,
                 the.data=climate)
new.list
```

```
$the.days
[1] "Monday"    "Tuesday"   "Wednesday" "Thursday"  "Friday"

$the.factors
[1] Small  Big    Big    Medium Medium Small  Medium Small  Small
Levels: Big Medium Small

$the.data
             Monday Tuesday Wednesday Thursday Friday
Temperatures     28      29        27       27     30
Rains             0       5         6        0      2
```

and use this to access elements by using $, as in data frames,

```
new.list$the.factors
```

```
[1] Small  Big    Big    Medium Medium Small  Medium Small  Small
Levels: Big Medium Small
```

```
new.list$the.data[2, 5]
```

```
[1] 2
```

Finally, the open structure of a list allows for an easy manipulation of data, for example, adding elements of a new kind to an existent list. We can do it by using the command c() and then use str() to know about the structure of the list,

```
new.list2 <- c(new.list, today="October 4")
str(new.list2)
```

```
List of 4
 $ the.days   : chr [1:5] "Monday" "Tuesday" "Wednesday" ...
 $ the.factors: Factor w/ 3 levels "Big","Medium",..: 3 1 1 2 ...
 $ the.data   : num [1:2, 1:5] 28 0 29 5 27 6 27 0 30 2
 ..- attr(*, "dimnames")=List of 2
 .. ..$ : chr [1:2] "Temperatures" "Rains"
 .. ..$ : chr [1:5] "Monday" "Tuesday" "Wednesday" ...
 $ today      : chr "October 4"
```

2.2.7 Exercises

Exercise 2.1 Take the vector rains from Sect. 2.2.2 and store in a new variable rains.monday the rainfall of Monday.

Exercise 2.2 Check whether it is true or false that 10 is less than 15 and 20 is more than 5.

Exercise 2.3 Guess the answer to the evaluation of this expression and check it out in **R**.

```
(((5 + 3)  ==  8 &  (7 < 4))  |  4 ^ 2 > 12)  & FALSE)  !=
FALSE
```

Exercise 2.4 Use vectors `rains` and `temperature` from Sect. 2.2.2 to

(a) subset the temperature of those days with any rainfall and store it in a new vector called `rainy.days`.
(b) subset those days with 27° and store it in a new variable called `t27`.

Exercise 2.5 Create a matrix 3×5 containing all consecutive numbers between 16 and 30 by columns.

Exercise 2.6 Add a new column to the matrix `total.climate` containing the total sums of each row; temperatures, rains, and winds. Add another column to `total.climate` with the average temperature, rain, and wind.

Exercise 2.7 Add to the `climate` matrix data for Saturday (temperature 31°, rain 0 L) and Sunday (temperature 26°, rain 14 L).

Exercise 2.8 We collect a survey about ten stocks position referred to last day, getting the sequence "U D D D U U U U D U," where "U" stands for "up" and "D" for "down." Create a factor vector with this information and specify that the value "up" is greater than "down".

Exercise 2.9 Gather the information of some colleagues from your class and create a data frame named `students` to store their data: name, age, degree, origin,... Try to find variables of all types, numerical, character, factors, and logical values.

(a) Using the `students` data frame, extract those students older than the average age, and those students of your same degree.
(b) Extract the origin country/region from those students older than the average.
(c) Sort the students data frame by using one of the quantitative variables.

Exercise 2.10 Use the data frame `CO2` from the `datasets` library to perform the following tasks.

(a) Have a look at the data frame to localize the different variables included and its type.
(b) Calculate the average of the CO_2 uptake rates in Quebec.
(c) Calculate the minimum and maximum concentrations of CO_2 for the chilled plants.
(d) For `Ms2` plant, how many levels of ambient CO_2 concentration were measured?
(e) For non-chilled plants, when the concentration is 500, what are the uptake levels?

Exercise 2.11 Create a list with different pieces of data from `students` data frame from Exercise 2.9. Give names to the elements in the list and access your name and age as sub-elements in the list.

Exercise 2.12 **R** provides abundant functions to all basic mathematical operations. Using the **R** help and the Internet, search how to round the following numbers: 3.2, 4.5, and 7.8. You should obtain the following results:

```
[1]  3
[1]  4
[1]  8
```

Try now to round to the second decimal on 3.273, 4.555, and 7.814:

```
[1]  3.27
[1]  4.55
[1]  7.81
```

2.3 Control Structures

Control structures are pieces of code that end up in different outcomes depending on the result of checking one or more logical conditions. It is an automated decision-making process where the program chooses to execute either one instruction or another, or even none of them, depending on the result of a previous part of the code. This previous output is passed as a new input and can control the rest of the program, the order of the execution and which parts are run or avoided from now on.

This is especially useful when the analyst wants to perform a task which is different depending on the variable or the situation. Imagine that in a particular situation the aim is to compute statistical measures in a data frame featuring a huge number of observations, but those measures are different depending on the columns (variables). If the variables are quantitative, it might be useful to compute mean and variance (see Sect. 5.1 for further information), whereas, if they are qualitative, a summary of frequencies for factor or logical variables is needed, as mean and variance cannot be calculated for them. Inside each group of variables we could be interested in performing an analogous analysis, or for each variable to do the same computation for each observation.

We distinguish two control structures. One is the conditional `if`, where certain instruction is executed upon the veracity of a condition, that acts as the control parameter for that part of the program to run. The other is the loop, which repeats an instruction several times and is implemented by two possible structures: `for`, when the repetition takes place a fixed number of times (which is the control parameter) and `while` when the iteration is not specified a priori but run as long as a condition is met (the control flag).

2.3.1 *Conditionals*

Conditionals are control structures where a logical condition plays the role of controlling whether or not some part of the code is executed. They are of great importance when analyzing different parts of data in a different way, depending on conditions. As an illustration, consider a study about ecological food consumer habits, where we gather survey answers of a population sample. Depending on the answer to the question "Have you ever bought ecological food?", the following questions and answers in the survey will be different and thus the dataset (which will be a data frame after preprocessing as it is explained in Chap. 3) will look different for individuals answering yes or no. Therefore, we are certainly interested in performing different analysis, calculations, statistics, or plots in these two situations. This is one among many situations where conditionals can be very useful.

The main conditional structure is given by if. We use the structure if to perform a task depending on a certain condition being true. The syntax is

```
if ("condition is satisfied") {
  "do something"
}
```

The condition has to be a piece of code whose evaluation in the console is TRUE or FALSE. Indeed, the if structure can naively run with these logical values instead of a condition, but the usual way is to place an expression which is eventually TRUE or FALSE.

For example, we begin with a value for the variable x, checks whether this value is greater than zero and, if so, prints the word "Positive"

```
x <- 3
if (x > 0) {
  print("Positive")
}
```

```
[1] "Positive"
```

If the initial value is negative, when checking the condition of being greater than zero we get FALSE, then the program does not enter inside the block between curly braces and, therefore, nothing is printed out

```
x <- -5
if (x > 0) {
  print("Positive")
}
```

Sometimes, we prefer to perform different tasks depending on whether the condition is satisfied or not. For this purpose we combine if with else. The syntax is

```
if ("condition is satisfied") {
  "do something"
} else {
  "otherwise do something else"
}
```

Going back to the previous example, we add the `else` structure where the word "Negative" is printed when the condition of being greater than zero is not satisfied. Now, if the initial value is negative, when checking the condition we get FALSE and the program enters into the second block between brackets, and prints "Negative".

```
x <- -8
if (x > 0) {
  print("Positive")
} else {
  print("Negative")
}
```

```
[1] "Negative"
```

If, otherwise, the initial value is positive, the program enters into the first block of code

```
x <- 6
if (x > 0) {
  print("Positive")
} else {
  print("Negative")
}
```

```
[1] "Positive"
```

But, what if the initial value is zero?

```
x <- 0
if (x > 0) {
  print("Positive")
} else {
  print("Negative")
}
```

```
[1] "Negative"
```

It makes complete sense from the point of view of the computer (which is, after all running, what we have programmed); zero is not greater than zero, then the first condition is FALSE and therefore the program executes the second block, hence "Negative" is printed out. Note that this does not mean that **R** is not working properly, it just means that our code was not precise enough and we labeled a set with a name that actually does not correspond to it, creating a potential problem for the rest of the code. Revisiting and checking the code thoroughly are a fundamental part of programming and should never be overlooked. To solve this counter-intuitive problem we can use several `else` blocks for checking multiple conditions and performing different tasks depending on the one satisfied.

The following code is a variation of the previous one that first checks the condition of being greater than zero, and enters into the first block to print "Positive" if it is true, as before. If not, there is a new condition to be satisfied, being less than zero. In such case the program runs into the second block and prints

"Negative". If neither the first nor the second conditions are satisfied, meaning x is neither positive nor negative, it enters into the last block and prints "Zero":

```
x <- 0
if (x > 0) {
  print ("Positive")
} else if (x < 0) {
  print ("Negative")
} else {
  print ("Zero")
}
```

```
[1] "Zero"
```

From the computational point of view, these structures are quite slow in **R** and can be outperformed by some shortcuts. The function ifelse() allows to execute two instructions upon the result of a logical condition in one line. The syntax for this is as follows

```
ifelse ("condition", "task if TRUE", "task if FALSE")
```

as in the example

```
x <- 9
ifelse (x > 0, "Positive", "Negative")
```

```
[1] "Positive"
```

The same condition can be verified multiple times by simply using a binary operator in a vector. For example, if we want to check whether numbers in a vector are lower than 5 we can compare the vector with 5 and get another vector of TRUE and FALSE entries depending on the result. We can use this logical vector as a vector of conditions for an ifelse() structure:

```
(1 : 10) < 5
```

```
[1]   TRUE   TRUE   TRUE   TRUE FALSE FALSE FALSE FALSE FALSE FALSE
```

```
ifelse ((1 : 10) < 5, "Fail", "Pass")
```

```
[1] "Fail" "Fail" "Fail" "Fail" "Pass" "Pass" "Pass" "Pass"
    "Pass" "Pass"
```

However, note the difference in the result with the for structure in Sect. 2.3.2 below. The output of an ifelse() structure is just one line, instead of one for each loop iteration.

2.3.2 *Loops*

Loops are one of the main structures in any programming language. They allow to perform the same task, or related ones, or even sequential ones a repeated number of times, upon certain control conditions specified in advance, or as the loop itself is being run.

We use the loop `for` to perform the same (or related or sequential) task a repeated and pre-specified number of times. The syntax is

```
for ("counter" in "vector of indices") {
  "do something"
}
```

Here the vector of indices is previously defined, this is, the code between brackets will run necessarily as many times as the length of the vector of indices. In each iteration of the code between brackets, the counter will run over the vector of indices taking each element at a time.

This example prints the value of the first 10 numbers

```
for (i in 1 : 10) {
  print(i)
}
```

```
[1]  1
[1]  2
[1]  3
[1]  4
[1]  5
[1]  6
[1]  7
[1]  8
[1]  9
[1]  10
```

In the example, the counter is i and the vector of indices is 1 : 10, this is, all consecutive integers between 1 and 10, both included. For i taking each value from 1 to 10, the instruction `print(i)` is executed, this is, printing the corresponding number. Note how the symbol [1] appears ten times, indicating that there are ten outputs, each containing one row. Therefore, we do not obtain a vector of the first 10 numbers, but a certain instruction, printing a number, which is run ten times.

The counter i runs over the set of indices,[50] but it also acts as a variable in the code between brackets. This will be very common, we are not just interested in performing a task a number of times, but want that task to be slightly different, depending on this (and maybe more) variable. Observe how i, as a variable, changes its value in each loop iteration and, after the loop is completed, it takes the last value of the index vector.

```
i
```

```
[1]  10
```

---

[50]Technically speaking, when using 1 : 10, **R** is internally doing a loop, so the previous code could be simplified to `print(1 : 10)` but it is valid as a first and easy example.

We can use more complicated structures for the loop recursion. The next example stores the cumulative sum of consecutive integers in a vector. First, we create an empty vector v with the function c ()

```
v <- c()
v
```

NULL

Then we set the variable s to 0 and, for each value of the counter i between 1 and 10, we do two things: first sum this i to the current s value and then assign this value of s to fill the following position in the vector v.

```
s <- 0
for (i in 1 : 10) {
  s <- s + i
  v[i] <- s
}
v
```

```
[1]  1  3  6 10 15 21 28 36 45 55
```

The role played by the variable s is to temporarily save the previous result (sum of all consecutive numbers) in each iteration of the loop. Note how we do not need to know the final size of vector v before starting the loop, we can always add new elements to our empty vector v <- c().

Of course, the vector over which the loop runs can be of very different nature. This loop starts with a vector of indices which is made of odd numbers

```
odd <- 2 * (1 : 10) - 1
for (i in odd) {
  print(i)
}
```

```
[1] 1
[1] 3
[1] 5
[1] 7
[1] 9
[1] 11
[1] 13
[1] 15
[1] 17
[1] 19
```

In this example, the set of indices is made of logical values and the instruction prints the comparison of TRUE with them

```
for (i in c(TRUE, FALSE)) {
  print(TRUE == i)
}
```

```
[1] TRUE
[1] FALSE
```

It is interesting to start combining both structures, the loop `for` with the conditional `if`. Imagine that we are grading a test and want to show the word *Fail* for those students not achieving at least 5 points, and *Pass* otherwise

```
for (i in 1 : 10) {
  if (i < 5) {
    print("Fail")
  } else {
    print("Pass")
  }
}
```

```
[1] "Fail"
[1] "Fail"
[1] "Fail"
[1] "Fail"
[1] "Pass"
[1] "Pass"
[1] "Pass"
[1] "Pass"
[1] "Pass"
[1] "Pass"
```

Finally, we use the control structure `while` to perform a task as long as certain condition remains true. It is crucial that the condition turns out to be false eventually, to prevent the loop from iterating forever. The syntax of this structure is given by

```
while ("condition holds") {
  "do something"
}
```

The following example starts with the variable `i` being 0. Then the instruction is to print a message saying that the value is less than 10, whenever this is true. At the end of each loop iteration we change the value of `i` by one unit more.

```
i <- 0
while (i < 10) {
  print(c(i,"is less than 10"))
  i <- i + 1
}
```

```
[1] "0"        "is less than 10"
[1] "1"        "is less than 10"
[1] "2"        "is less than 10"
[1] "3"        "is less than 10"
[1] "4"        "is less than 10"
[1] "5"        "is less than 10"
[1] "6"        "is less than 10"
[1] "7"        "is less than 10"
[1] "8"        "is less than 10"
[1] "9"        "is less than 10"
```

Note how the last iteration corresponds to 9, the last value which is actually less than 10. After printing `"9"` `"is less than 10"`, `i` happens to be 10 and when

checking if 10 is less than 10 we get FALSE, therefore the program exits this loop.
The variable i, after finishing the loop, takes the last value assigned

```
i
```

```
[1] 10
```

If we forget to change the value of the parameter with the line i <- i+1 and,
hence, change the logical condition to be checked, we step into an infinite loop as
in this example (Don't try it! Otherwise, be prepared to use the stop button in the
console!)

```
j <- 1
while (j < 5) {
  print(c(j, "is less than 5"))
}
```

where j is never different than 1 and that will display the following output repeated
ad infinitum

```
[1] "1"               "is less than 5"
```

2.3.3   Exercises

Exercise 2.13 Create an if - else structure which prints 0, 1, 2 depending on
the remainder of the division of a number by 3.

Exercise 2.14 Use the loop for to perform the following tasks:

(a) Show all consecutive numbers between 50 and 65 by using a loop for.
(b) Store in a vector called multiples.three all multiples of 3 between 1 and
     30.
(c) Store in a vector called powers the first fifty powers of 2.

Exercise 2.15 Compute the first 10 multiples of 3. Print "Below 15" for those
numbers below 15 and "Above 15" otherwise. Use a loop for and the conditional if. Repeat again this exercise using the function ifelse().

Exercise 2.16 Fibonacci sequence is defined by the recursion $x_n = x_{n-1} + x_{n-2}$,
where each element is given by the sum of the previous two. Use a loop for to
create the first 100 terms of Fibonacci sequence.

Exercise 2.17 Create a vector called consecutive, of 20 consecutive integers
between 15 and 34. Create another vector of the same length and a loop to
print "Odd" for positions where the corresponding element of consecutive is
"Odd", and to print "Even" otherwise.

Exercise 2.18 Study which numbers below 100 are divisible by 7. Use the control
structure while to output a list checking whether each number is divisible by 7 or
not.

2.4 Functions and Operators

Functions are the core object in mathematics. They are the main tool to perform certain tasks that vary upon initial conditions. A function is similar to a machine that receives one (or several) inputs and returns one (or more) outputs based on computations and operations with the inputs.

Mathematically speaking, a function $f(x)$ is a way to assign a value to each given value x. The function $f(x) = \exp(x) + 1$ consists of

- f is the name of the function. Traditionally in mathematics f usually represents a function (f coming from the name function) and, in case there are more functions, they are represented by g, h, etc.
- x is the argument or variable. It takes different values.
- $\exp(x) + 1$ is the function expression or body function. Indicates how to compute the function value, depending on the value of the variable.

The expression $f(x) = \exp(x) + 1$ can be read as *the function $f(x)$ takes, for each particular value of x, the result of substituting the expression* $\exp(x) + 1$ *by the x-value.* For example, when x takes the value 0, $f(0) = 2$, because $\exp(0) + 1 = 2$.

In **R** functions play the same role, although the argument x can be made of very different things, as well as the output $f(x)$. Single variables (numerical, character, or logical), matrices, data frames, lists (even outputs of other functions!) can be placed as inputs of a function. All kinds of structures can be outputs as well. The plethora of possibilities will give rise to very different body functions.

There are many functions implemented by default as part of the **R** packages. We have already visited plenty of them, but other examples are mean() for taking the average of all elements in an object, sum() for the sum of several quantities, sqrt() to calculate the square root of a number, log() and exp() to take logarithms or exponentials, trigonometric functions like the sine, cosine, and tangent of an angle with sin(), cos(), and tan() or logical functions such as is.logical() for checking whether the expression between parenthesis can be evaluated logically.[51] All of these are functions of a single input

```
mean(c(2, 5, 7))
```

```
[1] 4.666667
```

```
mean(matrix(1 : 4))
```

```
[1] 2.5
```

---

[51]Note that, in the example, the logical evaluation of the expression 3!=3 is FALSE, whereas being or not a logical expression is TRUE. Try the command is.logical("Hello") to see the difference.

```
sum(matrix(1 : 4))
```

```
[1] 10
```

```
sqrt(2)
```

```
[1] 1.414214
```

```
log(34)
```

```
[1] 3.526361
```

```
exp(0)
```

```
[1] 1
```

```
cos(0)
```

```
[1] 1
```

```
is.logical(3 != 3)
```

```
[1] TRUE
```

As we can see, in the previous examples, functions perform just one operation based on a single input, returning the result as an output with one line of code. More elaborated functions contain more than one element as an input. For example, in Sects. 2.2.3 and 2.2.6, we created matrices and lists depending on several arguments

```
matrix(3 : 8, nrow=3)
```

```
     [,1] [,2]
[1,]    3    6
[2,]    4    7
[3,]    5    8
```

In this example, we have used just two arguments (from all the possible ones) which are a vector and an integer number. And we have omitted more arguments which could have been used. This will be very common: functions allow to use several features; however, they are prepared to avoid some of them which are then understood to take a default value.

In data analysis, as a quantitative discipline, functions will play the same central role. Essentially, everything we do in programming is done by a function, and these functions can be either already in the programming language we are using, or we have to create them. **R** contains a huge repertoire of built-in commands developed for the main purposes in current industry and research. However, it is almost impossible that everything we need for a particular project is already created for us, and even when it is, we need to customize those commands for the specific analysis we are carrying out. A good example for these situations can be found in Sect. 5.1.2 where we develop functions to compute the population variance var.p and standard deviation sd.p because the ones included in the stats package are the unbiased variance and standard deviation.

We will learn how to create our own functions. The syntax is

```
function.name <- function(argument1, argument2,...) {
   "body function"
}
```

The name of the function, function.name, can be everything except for a *reserved word*, those words which are already used for something else in the programming language. The number of arguments can vary from none to many. The body function contains a series of instructions to execute depending on values taken by the arguments. Evaluation of last instruction is returned as output.

For example, suppose that some computation requires you to find the value of $f(x) = \frac{x}{1-x}$ repeatedly. Then we can write a function to do this

```
f <- function(x) {
   x / (1 - x)   # the output is the evaluation of last line
}
```

Note how there is a new element in the environment, in the category of functions, containing f. The name of the argument (in the example, x) is also a matter of personal choice. But the same name must be used also inside the body of the function.

This sort of *long* structure will be always necessary when the body of the function has more than one line. Otherwise we can use the shortcut

```
f <- function(x) x / (1 - x)
```

By default, a function returns the value of the very last line executed, therefore, one does not have to mention what to return explicitly. But the following form is also allowed in case we are interested in explicitly stating the output

```
f <- function(x) {
   return(x / (1 - x))
}
```

Once the function f is defined, it is stored in **R** workspace as a new function, and it can be used as follows

```
f(2)
```

```
[1] -2
```

```
y <- 4
f(y)
```

```
[1] -1.333333
```

```
f(2 * y)
```

```
[1] -1.142857
```

The same way we use functions of two variables in mathematics we do it in **R**. Imagine that we want to reproduce the function $g(x, y) = x^2 - \frac{y}{5}$, and compute various values

```
g <- function(x,y) {
    x ^ 2 - y / 5
}
```

```
g(1, 2)
```

```
[1] 0.6
```

```
g(2, 1)
```

```
[1] 3.8
```

We can evaluate functions in vectors. The result is another vector made of the evaluations on each element. The following code computes the exponential of the first ten integers, and then calculates the values of function f() over three numbers, 1,[52] 4 and 7

```
exp(1 : 10)
f(c(1, 4, 7))
```

```
[1]    2.718282    7.389056   20.085537   54.598150  148.413159
[6]  403.428793 1096.633158 2980.957987 8103.083928 22026.465795
[1]      Inf -1.333333 -1.166667
```

As we saw with the function matrix(), passing arguments is optional and, if no arguments are passed, the default values are used. The general way to write a function with default arguments is

```
funtion.name <- function(name.argument1=default.value1,
                         name.argument2=default.value2,...) {
    "body function"
}
```

We now explore how to create more evolved functions with all the aforementioned features. We want to design a chunk of code that creates either a vector or a matrix filled with the first integer numbers depending on the arguments that are passed upon the user call.[53] When the analyst introduces a number as the only argument, then a vector of the first consecutive numbers is ordered to the computer, whereas if another argument is passed for the number of rows, then the matrix is created.

The following code creates the desired function. There are three variables: a indicates the amount of elements for the matrix or vector, b tells the number of

---

[52]Observe that $f(1)$ is undefined, because we are dividing by zero. Despite this, **R** outputs Inf recovering the limits of f when x approaches 0.

[53]A richer function is already implemented in the **R** base library under the name mat.or.vec().

rows if the desired output is a matrix, and the logical `flag` will help the computer decide between matrices or vectors. Unless otherwise stated, the program will create a vector of numbers from 1 to a. If we switch `flag=TRUE`, a matrix containing integers from 1 to a, placed in b rows (where b=2 by default) is displayed.

```
mat.vec <- function(a, b=2, flag=FALSE){
    if (flag) {
      matrix(1 : a, nrow=b)
    } else {
      1 : a
    }
}
```

The purpose of setting values by default is that functions in programming usually have many different utilities. Instead of creating tons of functions for each particular situation, the idea is to reduce the number by plugging more arguments that allow a wide range of circumstances but which are disregarded if they are not needed. Once the number of arguments is considerable, a nice way to avoid error messages and stopping when running code is to set values by default; if we do not remember all arguments in a function, it will still run with those values.

We now try some evaluations of the function we just created. One thing to keep in mind is that we should give values to the function in the same order the variables are in the definition, then the first value corresponds to the first variable and so on. It is also possible to alter the order by passing them with the variable name, in that case switching the order is allowed but this practice is not recommended. First, simply create a vector of the first 5 integers, by giving the value of the variable a and leave `mat.vec()` to take the defaulted values for the other two variables

```
mat.vec(5)
```

```
[1]  1 2 3 4 5
```

Now, suppose we want a matrix with the first 6 numbers displayed in 3 rows, in which case we set a to the value 6, b changes to 3 and `flag` to TRUE, to actually have a matrix

```
mat.vec(6, 3, TRUE)
```

```
     [,1]  [,2]
[1,]    1     4
[2,]    2     5
[3,]    3     6
```

To illustrate the case where the order is changed (even though it is not recommended), the same can be achieved stating the variable names

```
mat.vec(flag=TRUE, a=6, b=3)
```

```
     [,1]  [,2]
[1,]    1     4
[2,]    2     5
[3,]    3     6
```

If we want a vector, it is enough not to call `flag`. Then, since it takes the value FALSE, argument b will be ignored even if passed; the vector is created ignoring the value of b

```
mat.vec(a=7, b=4)
```

```
[1] 1 2 3 4 5 6 7
```

If the matrix we try to build has a number of elements which cannot be divided by the number of rows, the matrix is still created and filled with the initial elements of the vector and a warning message is printed, as in Sect. 2.2.3

```
mat.vec(a=7, flag=TRUE)
```

```
     [,1] [,2] [,3] [,4]
[1,]   1    3    5    7
[2,]   2    4    6    1
```

```
Warning message:
In matrix(1:a, nrow = b) :
data length [7] is not a sub-multiple or multiple of the number
of rows [2]
```

The following example tries to create an element selector for vectors. Of course, this is already implemented in **R** with the syntax `vector[element]` but it is still instructional. It has two variables, v for the vector and pos for the position, the latter being defaulted to the first position, unless otherwise indicated. First, the program checks whether or not v is a vector with the command `is.vector` which returns TRUE or FALSE. In the first case, the selected element is given as an output. In the second case, the command `stop` will show an error message and the program will stop running, independently of the remaining code to be executed.

```
vector.selector <- function(v, pos=1){
  if (is.vector(v)) {
    return(v[pos])
  } else {
    stop("Variable v is not a vector")
  }
}
```

It is worth noting that, in this body function, the command `return()` is not necessary because it is the last line of the body function (in the case v is actually a vector). Nevertheless, it helps to immediately recognize the output in a larger function.[54]

---

[54]The computational advantages and disadvantages of using or not `return()` are beyond the scope of this book.

The first example selects the first element of the vector (3, 4, 5, 6, 7, 8)

```
vector.selector(3 : 8)
```

```
[1]  3
```

We can select a different position in a vector, by setting `pos` to a different value

```
vector.selector(c(1, 4, 5, 8, 2, 4, 3, 5), pos=6)
```

```
[1]  4
```

If the variable `v` is not a vector, for example a matrix, the error message is returned

```
vector.selector(matrix(1 : 9))
```

```
Error in vector.selector(matrix(1 : 9)) : Variable v is not a
vector
```

An **R** function is allowed to return only a single **R** object. To compute functions with more than one element as an output we can use lists. Suppose that we want to write a function that finds three things: length, total sum, and mean of a vector. Since the function is returning three different pieces of information we should use lists to store the output of this function.

```
items <- function(x) list(len=length(x),total=sum(x),
                          mean=mean(x))
```

Now we will use this over some concrete data. The output is a list where each element has its name, therefore can be accessed by the corresponding name with the $ symbol.

```
data <- 1 : 10
result <- items(data)
result
```

```
$len
[1]  10
```

```
$total
[1]  55
```

```
$mean
[1]  5.5
```

```
names(result)
```

```
[1] "len"    "total" "mean"
```

```
result$len
```

```
[1]  10
```

```
result$tot
```

```
[1] 55
```

```
result$mean
```

```
[1] 5.5
```

Sometimes you may be curious as to how a certain function is implemented. A simple way to peep inside the definition of a function is to type the name of the function at the console and the entire body of the function will be listed on screen.

```
log
```

```
function (x, base=exp(1))   .Primitive("log")
```

However, the output will often be overwhelmingly long (and somewhat cryptic to this level), even in simple cases as the mean.

```
mean
```

```
function (x, ...)
UseMethod("mean")
<bytecode: 0x1036b6d20>
<environment: namespace:base>
```

There are situations which require to apply the same function to more than one element. In these situations we can use the command lapply(list, function) which applies the function to all elements in the vector or the list. The command sapply() is a friendlier version of lapply() allowing to simplify the output to a vector or a matrix instead of a list.

```
salutation <- function(x) print("Hello")
# Note that this output does not depend on the value of x
output <- sapply(1 : 5, salutation)
```

```
[1] "Hello"
[1] "Hello"
[1] "Hello"
[1] "Hello"
[1] "Hello"
```

or we can even abbreviate it to

```
output <- sapply(1 : 5, function(x) print("Hello"))
```

for one line functions.

The data frame cars in the datasets library contains the speed of 50 cars and the distances taken to stop. Imagine that we want to compute the arithmetic mean of each variable, `speed` and `dist`. This can be done by interpreting the data frame as a list of 2 variable vectors, and `lapply()` computing the mean of each one:

```
lapply(cars,mean)
```

```
$speed
[1] 15.4
```

```
$dist
[1] 42.98
```

Or, alternatively, with `sapply()`,

```
sapply(cars, mean)
```

```
speed   dist
15.40  42.98
```

where we see how the output of `lapply()` is a list and the one of `sapply()` is a vector.

2.4.1 Exercises

Exercise 2.19 Make up two functions.

(a) Create a vector containing the names of the seven days of the week. Write a function that, given a number from 1 to 7, prints the name of the corresponding day.

(b) Write a function that, given a number from 1 to 12, prints the name of the corresponding month of the year.

Exercise 2.20 Write a function named `myfun` that computes x+2y/3. Use it to compute the expression $2 + \frac{2 \cdot 3}{3}$.

Exercise 2.21 Consider the function

```
g <- function(x, y) x ^ 2 - y / 5
```

Now try out the following and explain the output

```
g(1 : 2, 2 : 3)
```

```
g(1 : 4, 1 : 5)
```

What are we doing?

Exercise 2.22 Write a function to calculate the values of the polynomial $x^3 - x^2 + 2x - 4$ for $x = 0, 1, 2, \ldots, 10$.

Exercise 2.23 Write a function returning a list of the sum, the product, the maximum, and the minimum of a vector of values.

Exercise 2.24 Given a list of *n* values, create a function to calculate the determinant of the possible square matrix filled with these *n* values. First, check if *n* is a square. If so, create a square matrix with the *n* elements by rows, otherwise return an error message. Then, calculate the determinant (use the `det()` function).

Exercise 2.25 Use the function `sapply()` to print the message "I am so happy to learn **R** with this book!" 25 times.

References

1. Allaire, J.J. Rstudio: Integrated development environment for r. In *The R User Conference, useR!*, page 14, Coventry, UK, 2011. University of Warwick.
2. Allen, F.E. The history of language processor technology in IBM. *IBM Journal of Research and Development*, 25(5):535–548, 1981.
3. Austrian, G. *Herman Hollerith, forgotten giant of information processing*. Columbia University Press, New York, USA, 1982.
4. Babbage, C. *Passages from the Life of a Philosopher*. Longman, Green, Longman, Roberts, and Green, London, UK, 1864.
5. Blass, A. and Gurevich, Y. Algorithms: A quest for absolute definitions. In *Current Trends in Theoretical Computer Science: The Challenge of the New Century Vol 1: Algorithms and Complexity Vol 2: Formal Models and Semantics*, pages 283–311. World Scientific, Singapur, 2004.
6. Böhm, C. Calculatrices digitales. Du déchiffrage de formules logico-mathématiques par la machine même dans la conception du programme. *Annali di Matematica Pura ed Applicata*, 37(1):175–217, 1954.
7. Cardelli, L. Type systems. *ACM Computing Surveys*, 28(1):263–264, 1996.
8. Chambers, J.M.S. *Programming with data: A guide to the S language*. Springer Science & Business Media, Berlin, Germany, 1998.
9. Conference on Data Systems Languages. Programming Language Committee. *CODASYL COBOL journal of development, 1968*. United States Dept. of Commerce, National Bureau of Standards, Maryland, USA, 1969.
10. Copeland, B.J. *The Essential Turing*. Clarendon Press, Oxford, UK, 2004.
11. Dobre, A.M., Caragea, N. and Alexandru, C.A. R versus Other Statistical Software. *Ovidius University Annals, Series Economic Sciences*, 13(1), 2013.
12. Dybvig, R.K. *The SCHEME programming language*. MIT Press, Massachusetts, USA, 2009.
13. Friedman, D.P., Wand, M. and Haynes, C.T. *Essentials of programming languages*. MIT Press, Massachusetts, USA, 2001.
14. Gunter, C.A. *Semantics of programming languages: structures and techniques*. MIT Press, Massachusetts, USA, 1992.
15. Harper, R. What, if anything, is a programming paradigm?, 2017.
16. Hornik, K. The comprehensive R archive network. *Wiley Interdisciplinary Reviews: Computational Statistics*, 4(4):394–398, 2012.
17. Hornik, K. R FAQ. https://CRAN.R-project.org/doc/FAQ/R-FAQ.html, 2018. [Online, accessed 2020-02-29].
18. Ihaka, R. R: lessons learned, directions for the future. In *Joint Statistical Meetings proceedings*, Virginia, USA, 2010. ASA.
19. Ihaka, R. and Gentleman, R. R: a language for data analysis and graphics. *Journal of computational and graphical statistics*, 5(3):299–314, 1996.
20. Iverson, K.E. Notation as a tool of thought. *Commun. ACM*, 23(8):444–465, 1980.
21. Knuth, D.E. *The art of computer programming*, volume 3. Pearson Education, London, UK, 1997.

22. Knuth, D.E. and Pardo, L.T. The early development of programming languages. In *A history of computing in the twentieth century*, pages 197–273. Elsevier, Amsterdam, Netherlands, 1980.
23. McCarthy, J. Recursive Functions of Symbolic Expressions and Their Computation by Machine, Part I. *Commun. ACM*, 3(4):184–195, 1960.
24. Menabrea, L.F. Notions sur la Machine Analytique de M. Charles Babbage. *Bibliothèque Universelle de Genève*, 41:352–376, 1842. Translated, with additional notes by Augusta Ada, Countess of Lovelace, as *Sketch of the Analytical Engine*.
25. Posselt, E.A. and Philadelphia Museum of Art. *The Jacquard Machine Analyzed and Explained: with an Appendix on the Preparation of Jacquard Cards*. Published under the auspices of the school [Pennsylvania museum and school of industrial art], Pennsylvania, USA, 1887.
26. Price, D.S. A history of calculating machines. *IEEE Micro*, 4(1):22–52, 1984.
27. Pugh, E.W. and Eugene Spafford Collection. *Building IBM: Shaping an Industry and Its Technology*. MIT Press, Massachusetts, USA, 1995.
28. R Core Team. *R: A Language and Environment for Statistical Computing*. Vienna, Austria, 2018.
29. Racine, J.S. RStudio: A platform-independent IDE for R and Sweave. *Journal of Applied Econometrics*, 27(1):167–172, 2012.
30. Rogers, H. and Rogers, H. *Theory of recursive functions and effective computability*, volume 5. McGraw-Hill, New York, USA, 1967.
31. RStudio Team. *RStudio: Integrated Development Environment for R*. Massachusetts, USA, 2015.
32. Slonneger, K. and Kurtz, B.L. *Formal syntax and semantics of programming languages*, volume 340. Addison-Wesley Reading, Massachusetts, USA, 1995.
33. Truesdell, L.E. *The development of punch card tabulation in the Bureau of the Census, 1890-1940; with outlines of actual tabulation programs*. U. S. Dept. of Commerce, Bureau of the Census Washington, Washington DC, USA, 1965.
34. Turing, A.M. On Computable Numbers, with an Application to the Entscheidungsproblem. *Proceedings of the London Mathematical Society*, 2(42):230–265, 1936.
35. Van Roy, P. and Haridi, S. *Concepts, Techniques, and Models of Computer Programming*. MIT Press, Massachusetts, USA, 2003.
36. Wickham, H. *R Packages: Organize, Test, Document, and Share Your Code*. O'Reilly Media, California, USA, 2015.
37. Winskel, G. *The formal semantics of programming languages: an introduction*. MIT Press, Massachusetts, USA, 1993.

Chapter 3
Databases in R

All what was considered in the previous chapter is of extreme importance, yet it is all part of what can be referred to as traditional programming and none of it was specifically oriented to data analysis. This chapter makes extensive use of the knowledge from the previous chapter but focuses on using those techniques and learning new ones in order to deal with large volumes of information. More specifically, we will overview everything that is needed to go from the *first contact* with new data to the moment when the analyst is ready to start the actual data analysis, which will be studied in future chapters by using adequate statistical learning methods.

First, we will go over how to import and export data from available *sources*. Accessing structured data can be straightforward, if it is already given as a file, or downloaded through APIs from an internet source. But there are many times where no such API is at hand, and web scraping is the way of getting pseudo structured data. Finally, preparing data to be understood and used by algorithms extracting conclusions is done through *preprocessing*, which will format data in such a way that it can be used by the analyst.

3.1 Data Sources, Importing, and Exporting

As it was discussed in the introduction, the world is now data driven. The amount of data generated and stored all over the world is enormous, and it is so in all fields, areas, and sectors that one could think of. However, accessing data is usually far from easy. There are many reasons why data is not shared. One of the most serious reasons is privacy and confidentiality. Data involving personal information is always protected somehow by law, and cannot be shared with others just because. However, most of the time the reason is even simpler. Gathering and storing data is a tedious and cumbersome task and once someone has done it, it has an overwhelming value

© Springer Nature Switzerland AG 2020
A. Zamora Saiz et al., *An Introduction to Data Analysis in R*, Use R!,
https://doi.org/10.1007/978-3-030-48997-7_3

which can be transformed into huge profit so, of course, no one is willing to share
such a valuable good.

Prior to any kind of processing, there are four aspects to consider, no matter what
kind of data an analyst is facing or what kind of analysis is going to be performed.

- **Obtainment**: As stated before, even though data is everywhere, obtaining data
  might range from something as easy as receiving a tabulated file with all the
  information to something advanced such as programming an automated script
  that gathers the information.
- **Reliability**: This, of course, is not new of modern data analysis, but can have
  a deep impact on it. The data source itself must be called into question. Are
  we obtaining the data from the original source? From a third party? When we
  generate the data ourselves, what might be going wrong? On top of an abstract
  discussion about the reliability, all data sources need to be deeply examined in
  order to verify how trustworthy they are and to check for possible inaccuracies.
- **Storage**: Sometimes, and every time more frequently, the amount of data is so
  much that it will not fit even on the largest hard drive. This is the first reason
  to embrace online solutions, where the advantages of working on the cloud
  overcome the problems of storage.
- **Processing**: Even when the volume is not so big and can be stored in a standard
  computer, it is frequently the case that RAM memory is overloaded. For instance,
  a 10 GB database in a single file can be stored in any computer hard drive but
  such a file cannot be entirely loaded in the RAM memory, since most computers
  do not usually have more than 8 GB of RAM. Two alternatives arise here. Some
  specialized software solutions are able to open the file sequentially, loading in
  RAM at every moment only what is strictly needed, or even capable of reading
  the file directly from the hard drive. Even though this is a very clever workaround,
  it slows down the loading and writing. The other alternative is of course cloud
  computing, and that is why it has become so important.

The data scientist needs to keep these in mind but once these aspects have been
dealt with, they will no longer affect the conclusions of the analysis. The fact that
a method is applied locally on a computer or using online services allowing for
distributed computing[1] affects deeply the efficiency and the way they should be
implemented.[2] Nonetheless, this does not affect the particular choice of an algorithm
for a given problem and what code to write in **R** no matter how its backbone
is programmed. In the same vein, once the authenticity of the source has been
ascertained the code to be executed is identical regardless of whether the final
conclusions can be claimed with more or less certainty.

---

[1]Distributed computing is a model in which components of a software system are located on
different networked computers, as in cloud computing.

[2]Researchers around the world keep improving the implementation of specific algorithms to take
full advantage of the very nature of distributed computation.

We will not focus on reliability of the data, as there are no specific patterns or rules for this. As with traditional sources of information, the analyst should be extremely critical here. When data is provided or acquired from someone else we have to ask ourselves if the source can be considered trustworthy. What might be new in modern data analysis is that most data are generated automatically, and that is a double-edged sword. On the one hand you are obtaining the data directly from the original source, with no human bias or errors. On the other hand, a sensor retrieving information might have been badly calibrated, or your online bot in charge of reading a web might have been offline for a while due to loss of internet connection. There are no shortcuts or checklists, only a deep examination of all that could have gone wrong.

However, we will focus on the other three aspects. An independent analyst will continuously face the problem of accessing data but even an analyst working for a firm with full access to all its files will find him/herself in the same situation as soon as there is the need to enhance the study with exogenous data from outside the company. Thus, we devote Sect. 3.2 to *scraping*, the art of obtaining data which is available to us, but not easily accessible.

Regarding storage and processing, when the database exceeds the capabilities of a personal computer, the only option is to search for online solutions. The main options as of the date of writing are services such as Amazon Web Services, Microsoft Azure, Google Cloud, IBM Cloud, Oracle, MongoDB, and Hadoop. All of them offer a variety of possibilities and powerful solutions. With respect to storage, the options range from very cheap plans where the file is stored but can be accessed a limited amount of times and only upon request (think of a big data base that was already studied and needs to be stored but is not going to be used soon) to more flexible options were the file can be accessed as if it were in your personal computer. As for computing, several options are also available. Some are focused on fast access to data, other services allow for parts of code to be easily uploaded and executed by "hiring" the services for a certain amount of time, at cheap prices. More proficient tools such as loading a full virtual machine are also available if the code needs of further interactions with other programs or services and everything needs to be uploaded as a whole.

Online options is a rich topic, with enough material for a whole book, but we do not dive into its details since the options, pricing plans, and the tools themselves are constantly changing. For further and updated information the authors recommend to visit the official webpages of each service.

Once these first considerations have been discussed, we will discuss how to import and export data. Remember that in Sect. 2.2.1 we explored the different types of variables that one can find. We had numbers, strings, dates, or more complicated ones such as factors, data frames, lists, etc. Note, however, that we have not said anything about saving our work. We could have created a useful table for a certain analysis but once RStudio is closed all information is lost.[3] We show how to import data (to a data frame that we create) and export (from a data frame that we already have), and to do so, we need first to introduce some concepts about file types.

3.1.1   Tabulated File Types

A tabulated file is a file where certain characters act as separation between contents. The most basic tabulation is a file where each piece of information inside certain row is separated by spaces and rows are separated by an end of line. This is convenient for many situations but it has a major problem, information with two words cannot be stored. For example, the information *orange shirt* will be stored as two pieces of information. Thus the following table for prices of clothing

          jean orange shirt jacket sock
          50   30   70   10

will be stored as

jean orange shirt jacket sock
50 30 70 10

loosing all tabulation and therefore read as

          jean orange shirt jacket sock
          50   30   70   10

making it look like there is an extra product (oranges) and scrambling all the values. Needless to say that storing that file in such way will make it useless. In order to solve the problem with spaces, values can be separated by commas. Now the table will be saved as

jean, orange shirt, jacket, sock
50, 30, 70, 10

and therefore read correctly

          jean orange shirt jacket socks
          50   30   70   10

---

[3]Technically speaking, RStudio automatically saves the session when closing so the work is actually preserved, but it is not stored as an external editable file, just as a snapshot of the session.

However, potentially new problems arise. If information contains commas, such as sentences or when using the comma as the decimal separator,[4] then the information will be saved incorrectly. Think of saving the numbers 3, 14 and 2, 71; they will be stored as 3, 14, 2, 71 and then read as four different integer numbers. To avoid this new problem the information can be separated using semicolons ; that appear really rarely in common sentences or data, but it is not free of risk. As a general rule, there are no shortcuts or rules to follow. The analyst needs to be vigilant of the separator; when importing data, the correct separator has to be used to load the table correctly and, when encoding, doing the right choice for the separator according to the specific data.

Tabulated files can be saved in many file formats. The most common and important file format for tabulated data is the *comma separated values format*, .csv in what follows. Even though the name might suggest that the separation character for these files are commas, the truth is all possible separators are admissible. Therefore, both when exporting and importing data we have to specify the separator.

3.1.2   Importing and Exporting

In order to *import* a .csv file we use the command read.csv(). That command is, as the ones in Sect. 2.4, a function that receives several inputs as arguments. We will focus in four of them, the first is where to read from, the second decides whether the stored table has header or not, then the separator is specified and finally the decimal separator. For example, if we want to read a file that has headers, that is separated by semicolons and using the dot as the decimal separator, the command is

```
our.table <- read.csv("myfilewithdata.csv", header=TRUE, sep=";",
            dec=".")
```

The table is read into the object our.table which is a data frame (see Sect. 2.2.5).

Likewise, the command write.csv() is used to *export* data frames into a .csv file. Similar arguments are passed to the function, specifying now whether the names of rows and columns should be included in the file. For example, if we want to write a table into a file that keeps the names of the columns, discards the row names, separated by commas and using the dot as the decimal separator, the command is

```
write.csv(our.table, "filetosave.csv", col.names=FALSE,
          row.names=TRUE, sep=";", dec=".")
```

where the first argument selects the object to save, the second one specifies the name of the file, then the information about names in columns are given and finally the separators are indicated as above.

---

[4]The comma is used as a decimal separator in many countries in the world, see https://en.wikipedia.org/w/index.php?title=Decimal_separator&oldid=932234568.

Even though `.csv` is by far the most used format, other file formats are out there and need to be handled. For files such as `.txt` or `.dat` use the functions `read.table()` or `read.delim()` with almost the same arguments as `read.csv()`. Files can be saved in those formats in a similar way but the authors recommend using `.csv`, as it is more standard and, therefore, easier to read somewhere else.

The commands discussed above are, however, not optimized for big data files. Both `read.csv()` and `write.csv()` will take large amounts of time to read and save such data tables. Luckily, there is an **R** package, `data.table`, that has developed functions with the specific purpose of loading and saving data in extremely faster times. The technicalities of why this package is so much better are beyond the scope of the book, but keeping in mind the efficiency in performance of the tools we use is fundamental.

We load the package `data.table` by `library(data.table)`[5] and then use `fread()` and `fwrite()` to read and write files, respectively.[6] There are only advantages when using these two functions; the loading and saving speed is extremely faster for large files; it is able to read from different formats, not only `.csv`; when reading, the function usually recognizes the separator without any help from the analyst, and also what kind of data is found in each variable; the table is not only opened as a `data.frame` but it is embedded with the `data.table` structure, an enhancement of data frames that will be explored later on.

In addition to tabulated files, the analyst will surely find *hierarchical files*, that is, files structured into different categories, that contain other sub-categories inside and so on, creating a hierarchy. The absolute standard for hierarchical files is the *JavaScript Object Notation* or `.json`. This file format stores the information as if it were a table of contents from a book, creating an easy to understand and intuitive file. However, that structure is not adequate for data analysis and it must be *parsed*, that is, converted into a data frame. There are several libraries for parsing from `.json` files in **R** from which we will use `jsonlite`.[7] In order to parse a `.json` file into a table, just type

```
library(jsonlite)
jsonastable <- as.data.frame(fromJSON("ourjsonfile.json"))
```

We cannot finish the discussion on file formats without mentioning *spreadsheets*. Spreadsheets are great and powerful tools for a variety of areas and are widely used in almost every business for accounting, descriptive statistics, basic databases, easy calculations, etc. Nonetheless, spreadsheets lack many of the features that a programming language provides in modern data analysis and also have severe

---

[5]Using different libraries will be a constant throughout the book. Whenever a new package is introduced it is understood it should be installed first, even if it is not specified. See Sect. 2.1.

[6]Here the f in `fread` and `fwrite` stands for fast.

[7]As of 2019 the community seems to agree that the package `jsonlite` is a bit better than `rjson` or `rjsonio`.

shortcomings. Microsoft Excel[8] is undoubtedly the most popular spreadsheet application. In the particular case of this spreadsheet, the most important limitations are speed of calculation and that there is a maximum for the number of rows and columns that an Excel file might have, 1,048,576 and 16,384, respectively. An Excel file, .xls or .xlsx, is a tabulated file format with many enhancements. The file preserves colors, graphics, formulae, formatting, and everything that a user can create inside Excel. That is, however, another drawback since all those features lack of any value for the data analysis to be carried out in **R** but contribute to a larger and less compatible file. Nevertheless, Excel spreadsheets are so common that importing from them should not be disregarded. Again, several libraries are available but the fastest one with less dependencies on other packages and most versatile is readxl. We conclude the section presenting an example where the second spreadsheet from a .xlsx is imported, and where we specify the type of data in each column as well as the presence of headers.

```
library(readxl)
data <- read_excel("file.xlsx", col_names=TRUE,
                   col_types=c("numeric", "numeric"), sheet=2)
```

Note that the first two arguments are rather similar to the ones before, but some new inputs appear, one is col_types which specifies what kind of data is present in every column read and the last one states what sheet to read, in case there are more than one in the file.

Finally, as it was mentioned in Sect. 2.2.5, **R** includes a variety of built-in datasets with the purpose of providing useful objects for examples or as a learning tool. Furthermore, plenty of packages add some extra datasets in relation with the usage of that library.

3.1.3 Exercises

Exercise 3.1 One of the datasets mentioned above which are included by default in **R** is cars, from the package datasets. Save the cars dataset into a .csv by means of both write.csv() and fwrite() making sure that the names of the variables are included in the file but the names of the rows are discarded. If no separator is specified for the tabulation of decimal punctuation, what characters are being used for that matter?

The following exercises make use of files that can be downloaded from the book data repository at https://github.com/DataAR/Data-Analysis-in-R/tree/master/datasets. These files feature certain information on worldwide flights. More on this data is commented in the next section.

---

[8]https://products.office.com/en/excel.

Exercise 3.2 Load `easy.csv` and `hard.csv` using both `read.csv()` and `fread()`. Compare the speed of both functions and the simplicity when dealing with a non-standard file.

Exercise 3.3 Load `flights.json` and `flights.xlsx` making sure that no information is lost.

3.2 Data Collection

According to Data Never Sleeps 6.0, an annual report on the state of the Internet performed by Domo Inc. [2], more than 90% of all data generated in human history was generated in the last 2 years. Just as an example, for every minute in 2018, more than 750,000 songs are played on Spotify, 3,877,140 Google searches are conducted, 97,222 h of video are streamed by Netflix, and 473,400 tweets are sent.

The previous section explored how to import data but, in all cases, it was assumed that the data to import was already stored in a file, with a specified format and given to the analyst. However, most of the time, data cannot be easily accessed by the user and even when it is reachable, it will possibly be encrypted or stored in a way that is not useful for data analysis. We need to distinguish different kinds of data regarding its structure [4]. Three main categories can be defined.

- **Fully Structured**: A dataset where the variables are clearly defined, which is tabulated and ready to use. One can think of this category as any file that can be transformed into a data frame effortlessly with the commands learnt in the previous section. This kind of structure is what one finds when using APIs (to be introduced below) and in open data sources, where a previous work on tabulating the data has been done.
- **Semi Structured**: Data where the variables are still clearly defined, but no previous formatting has been done. This category may be thought of as data that can be transformed into a data frame after a (possibly hard) preprocessing. An archetypal example for this kind of data are webpages, where contents tend to be organized and can be easily followed through links, headers, captions, etc., but where retrieving all data at once in a tabulated file is usually far from easy. Much more on this will be said below.
- **Unstructured**: In this case the choice of variables is not clear. Typical examples are images, music, or unprocessed text. It is clear that all of those might be of extreme value but we do not even know how to open that kind of information with **R**. Even when the possibility of opening them from **R** exists, an extra layer of preprocessing is required to transform that new structured object into a data frame suitable for analysis.

Finding the right way to process unstructured data involves very deep knowledge in mathematics and is beyond the scope of this book, therefore we will only focus on the other two.

3.2.1 Data Repositories

The easiest way of accessing data is when it is provided already as one of the standard files explained in Sect. 3.1. Open databases have become popular lately: governments, public institutions, and other organizations release data in order to inform society, stress situations in their agendas, or just with the purpose of showing a clean image of transparency within the administration. Some widely known examples are The World Data Bank,[9] Eurostat[10], or the U.S. Government's Open Data.[11]

Sometimes, databases are shared just as an educational tool, such as the UCI Machine Learning Repository.[12] The popularity of this webpage is such that a package featuring some of the most famous datasets is developed for **R**. In order to load one of the sets, such as the `Glass` dataset, just type on your **R** console

```
library(mlbench)
data(Glass)
```

Often the reason to share databases is to test new algorithms, see, for example, the list proposed by Wikipedia.[13]

Finally, databases are shared as a commercial product. This solution is particularly frequent in economics where it has been clear for decades that the potential profit of having access to data is worth it even if payment is required. Some famous solutions in the field of economics are Bloomberg[14] or Global Financial Data.[15] Other services such as Euromonitor[16] can be found outside finance.

All these databases have exporting tools implemented so data can be downloaded in a standard format that can be opened with the importing functions studied in Sect. 3.1.

3.2.2 APIs

The process of surfing to a webpage, following some links and then downloading a dataset, is very inefficient in various aspects. From the point of view of the user, it is very difficult to automate and thus time consuming, especially if the process is to be repeated on a regular basis. On the other hand, if a platform is sharing a 1 MB

---

[9]http://www.worldbank.org.

[10]https://ec.europa.eu/eurostat.

[11]https://www.data.gov.

[12]https://archive.ics.uci.edu/ml/index.php.

[13]https://en.wikipedia.org/wiki/List_of_datasets_for_machine_learning_research.

[14]https://www.bloomberg.com.

[15]https://www.globalfinancialdata.com.

[16]https://www.euromonitor.com.

file every minute, they are not willing to make the user access that file through several links, each of them adding some MBs to each file transmission that will unnecessarily overload the servers.

In order to offer direct access to information, Application Programming Interfaces (*APIs* in what follows) have become popular. APIs are communication protocols with a clear structure that allow efficient transmission of information. APIs have additional advantages, for example, access to data can be easily controlled or restricted by limiting the amount of MBs or the amount of connections in a certain amount of time, or checking for credentials if needed. Another benefit shows up when the amount of data is so enormous that it just cannot be accessed in a single file to be downloaded, hence fragmenting it through these protocols becomes of great importance.

3.2.2.1 REST APIs

Every API has its own documentation where fully detailed explanations on how to use it are given. Some of them are conceived for specific programming languages such as Python or PHP[17] so learning how to use them implies at least some knowledge on various languages. On top of that, the world of APIs is an ever changing one, as new platforms are constantly springing and stable ones keep updating their protocols, therefore, a full guide on the use of every possible API is not feasible. Nonetheless, the most common solution is Representational State Transfer APIs (*REST APIs*),[18] a kind of API where all requests are handled strictly through URLs[19] and data is provided in standard structured formats. The main idea when using REST APIs is that all information will be accessible through several links that have an understandable structure. Tweaking some parameters of each link will change the information that is requested or some other features, such as the period of time to analyze or how the output is presented. As before, every REST API has a particular way of structuring the URLs to access the data, so a detailed procedure for all of them is not possible. However, patterns are repeated in all platforms so two main steps to keep in mind should always be considered:

- First, read the documentation to find the links that feature the desired information.
- Then, access those links from **R** and convert the information into a data frame if need be.

There is not much more to say about the first part, as it just consists on reading the documentation in detail to find the relevant links. The second part, though, is more methodical since, once we have a URL, the way of accessing it from **R** is always

---

[17]PHP stands for Personal Hypertext Processor, and is a language for webpages.

[18]Often the so-called RESTful APIs for the advantages and easiness when using them.

[19]URL stands for Uniform Resource Locator and corresponds to what is usually called *web address* or *link*.

the same. We explore this two-step process through some examples to enlighten the details.

3.2.2.2 OpenSky Network Example

The OpenSky Network is a non-profit association with the aim of improving the security and efficiency of air traffic that shares data through a REST API. Even though they share the data for free, some requests can only be pulled by authenticated users and some are completely public. For this example, we will get the state of all flights currently airborne.[20]

The first step when dealing with a REST API is to read the documentation.[21] Parts of the documentation might be skipped if the commands in those parts are not useful for the study to be carried out, but the terms of usage, limitations, and best practices should always be read in detail and agreed. Inside the documentation the analyst should look for a link giving access to the required data. It is a common practice with APIs to offer query examples as the main way of teaching how to use it, assuming that the analyst will explore the particular tweaks in the parameters of the URL to access the desired information. In this case, in order to retrieve data of the flight states, the documentation gives URL examples for several requests:

- Example query with time and aircraft: https://opensky-network.org/api/states/all?time=1458564121&icao24=3c6444

Such a request will provide the information about the state of plane with code 3c6444 at time 1458564121. It is also explained in the documentation that a query of all states is conducted in the following way skipping the part of the link referring to time and code of plane. Thus, since the target is downloading the states of all flights currently airborne we just need the URL https://opensky-network.org/api/states/all. Just by opening that link in an internet browser, the list of all states at that certain moment is displayed on the screen as a .json file. If the link is refreshed, the data will be updated, so that the current snapshot of all the information can be retrieved just with the use of this link as it is always the case for REST APIs.

The display is an utter gibberish as it is conceived as a minimalist way of efficiently accessing the data, not as a friendly looking interface. Luckily, our true purpose is not to extract conclusions by displaying these data on screen, but to open it from **R** and provide it with an adequate structure, desirably a data frame. For that, open RStudio and create a new .Rmd file (see Sect. 2.1). Since the response to requests is a .json file and our intention is to convert it into a data frame, we introduce a first block to load the required packages

```
library(jsonlite)
```

---

[20] All data available in this subsection has been kindly provided by The OpenSky Network, https://opensky-network.org. See also the original OpenSky paper [10].

[21] For this particular case, see https://opensky-network.org/apidoc/rest.html.

Then, calling the URL and importing the data could hardly be easier, just introduce a second portion as follows:

```
url <- "https://opensky-network.org/api/states/all"
flights <- as.data.frame(fromJSON(URLencode(url)))
```

The first line stores the URL into the variable `url` as a string. In the second line, the command `fromJSON()` is prepared to read from a `.json` file, so the command `URLenconde()` is added in order to tell **R** that the argument in this case is a JSON, which happens to be embedded inside a URL. At this point, the resulting object of doing `fromJSON(URLencode(url))` is a list of two items, `$time` and `$flights`. In order to have the more convenient data frame, the information is converted using `as.data.frame()` and stored as the variable `flights`.

Note that even though the importation from the internet is finished, the dataset still needs some processing and preparation prior to starting the analysis. What is each variable? What is the date format? What happens with `NA`s? All these questions are discussed in Sect. 3.3.

3.2.2.3 TheSportsDB Example

For the second example we will explore TheSportsDB,[22] according to their own description, *"an open, crowd sourced database of Sports metadata"*. In their documentation[23] we find that an *API KEY* is needed. An API KEY is a user credential to identify who is connecting to the database and their use is extremely common to prevent unauthenticated connections even in open databases that have no problem with sharing the data but want to control the traffic. The usage is usually very simple, a portion of the URL has to be substituted by the users KEY. Fortunately, TheSportsDB generously allows for public connections using the key `1` but emphasizes that this key should only be used for test purposes, as we do now.

The task now is to obtain all the information about matches in "La Liga," the Spanish soccer league, in the season 2017–2018. Reading through the documentation, we find again many examples to access the information. To retrieve the desired information we use the following bit:

- All events in specific league by season: https://www.thesportsdb.com/api/v1/json/1/eventsseason.php?id=4328&s=1415

As before, the first part of the URL, before the question mark, is general and writing `id=4328&s=1415` forces to look for a specific match in one season. Note that part of the url reads "`/1/`," that is, precisely the part that should be replaced by

---

[22]https://www.thesportsdb.com.

[23]https://www.thesportsdb.com/api.php.

an API KEY if we had one. As before, it is enough to open RStudio and follow the same commands

```
sport.url <- "https://www.thesportsdb.com/api/v1/json/1/
           eventsseason.php?id=4328&s=1415"
sports <- as.data.frame(fromJSON(URLencode(sport.url)))
```

The data frame `sports` looks amazing, not only lists all matches but it features over 50 columns of information ranging from date or result to scorers, lineups, or yellow cards. However, the obtained information is about the Premier League (the professional soccer league in England) in the season 2014–2015, not the one we are looking for. It is very often the case that the documentation of REST APIs is written through examples, leaving further interpretation to the reader, as all variations of the requests can be sorted out by small manipulations of the URL. In this case, note the text "`id=4328&s=1415`" at the end of the URL. It should be understood that `id` identifies the league (being `4328` the code for Premier League), `s` is for the season, and `1415` stands for 2014–2015. It is then clear that the second part should be `s=1718` for our example but what is the identifier for "La Liga"? Going back to the documentation, the following line can surely help us:

- List all Leagues in a country: https://www.thesportsdb.com/api/v1/json/1/search_all_leagues.php?c=England

and introducing the following in the console prompt:

```
leagues.url <- "https://www.thesportsdb.com/api/v1/
            json/1/search_all_leagues.php?c=Spain"
leagues <- as.data.frame(fromJSON(URLencode(leagues.url)))
```

we find out that the identifier for "La Liga" is `4335`, so finally write

```
sport.url <- "https://www.thesportsdb.com/api/v1/
            json/1/eventsseason.php?id=4335&s=1718"
sports <- as.data.frame(fromJSON(URLencode(sport.url)))
```

to obtain the desired frame.

This example enlightens how the analyst should read the documentation and pull the thread to find the adequate command to the purpose of the research. One further warning of great importance should be mentioned: both examples feature the command `as.data.frame(fromJSON(URLencode()))`. This structure will be fine with most REST APIs but it is not a perfect recipe. It is very infrequent that a REST API does not provide a `.json` file as one of the possible responses but sometimes the structure of the file is such that conversion to `data.frame` cannot be performed "out of the box." We end the subsection with an example of such a case.

Still working with TheSportsDB, try to retrieve all soccer live scores. According to the documentation and proceeding as before we would write

```
matches.url <- "https://www.thesportsdb.com/api/v1/
             json/1/latestsoccer.php"
matches <- as.data.frame(fromJSON(URLencode(matches.url)))
```

```
Error in (function (..., row.names = NULL, check.rows = FALSE,
check.names = TRUE, : arguments imply differing number of rows:
1, 0
```

Clearly, something went wrong. When such kind of things happen, it is recommended to go step by step to see what is going on. Just by calling

```
matches <- fromJSON(URLencode(matches.url))
matches
```

we obtain

```
$teams
$teams$Match
```

and the whole table. Remember that `.json` files are hierarchical files. Here, there is a hierarchy, `teams`, with only one sub-hierarchy, `Matches`, containing the whole table. The command `as.data.frame()` is not able to understand such structure on its own, so we need to manually specify to look for the dataset and convert it with the following command:

```
live.matches <- as.data.frame(fromJSON(
                             URLencode(matches.url))$teams$Match)
live.matches
```

working now perfectly.[24]

There is no shortcut for this kind of issues. The researcher should read, try, retry, and explore in order to solve the particularities of each call of an API. Sometimes, requests can get so complicated that specific **R** packages exist to deal with those APIs (see Exercise 3.6 below). As it was the case with the example of the flights, the retrieved dataset still needs some preprocessing before it can be analyzed using the tools in Chap. 5.

3.2.2.4   Quandl

For the last example, we will explore Quandl,[25] a popular platform for financial data that delivers a powerful yet simple API. In the documentation[26] a new particularity is found, an **R** implementation is provided as well as the REST API. This means that Quandl can be used in two ways in **R**, by using the library `Quandl` that developers created, or in the standard way of accessing links that we have been training. It is important to clarify that when a platform provides these two ways of accessing data, both approaches can accomplish the same results. The difference lays in performance or in how to write the commands. Usually, the **R** library will

---

[24]Make sure to run this code while matches are being played, otherwise, an empty file will be generated.

[25]https://www.quandl.com/.

[26]https://www.quandl.com/tools/api.

feature functions that are styled as specific commands in **R** that directly provide the desired data frame, whereas the URL approach will possibly require a longer code for the same output, but that code is standard, similar to what has been studied so far, and more easily exportable to other platforms. Both approaches have advantages but for the sake of consistency we will keep working with the links instead of using the specific package.

The task for this last example is to retrieve the closing values for the security daily of Google since January 2015. Reading through the documentation we find, again as particular examples, that

- This simple call gets a historical time-series of AAPLs stock price in CSV format: https://www.quandl.com/api/v3/datasets/WIKI/AAPL.csv
- Set start and end dates like this: https://www.quandl.com/api/v3/datasets/OPEC/ORB.csv?start_date=2003-01-01&end_date=2003-03-06

Note that the code queried in the first link is AAPL which corresponds to Apple's security daily. In order to look for Google's information we need to refer to its ticker GOOG.[27] In the second link, the last part of the link includes the information about date restrictions. As before, it is enough to open RStudio, but now recall from the URL that the file is given as a .csv so there is no need to load the package jsonlite. Instead, we use the function fread(),

```
library(data.table)
quandl.url <- "https://www.quandl.com/api/v3/datasets/
              WIKI/GOOG.csv?start_date=2015-01-01"
fread(URLencode(quandl.url))
```

A data frame containing the close value, and many other variables, is returned. Revisiting Quandl's documentation, plenty of options are found, some of them focused on basic preprocessing. We do not develop this part of the API as the whole Sect. 3.3 is devoted to this topic in detail, including advanced commands and proficient techniques that outperform the methods that are sometimes implemented in APIs.

Among data driven companies it is hard to find one that does not have an API, as their business lays precisely in dealing with information. However, obtaining credentials for them is becoming more and more difficult by the day, as platforms are reluctant to share data in exchange of unclear return. Twitter has always limited the amount of requests per day but as of 2018[28] the credential system to access the API[29] and the requests limits[30] are getting more restrictive. Google had an open

---

[27]Tickers are not provided in the API documentation, but can be easily found in www.quandl.com or googled.

[28]https://twittercommunity.com/t/details-and-what-to-expect-from-the-api-deprecations-this-week-on-august-16-2018/110746.

[29]https://blog.twitter.com/developer/en_us/topics/tools/2018/new-developer-requirements-to-protect-our-platform.html.

[30]https://twittercommunity.com/t/new-post-endpoint-rate-limit-enforcement-begins-today/115355.

free-to-use API[31] for Google Maps but it is commercial since 2018 and the billing options have been criticized;[32] we will use its free and limited version in Sect. 4.4. Possibly, the most extreme case is that of the Instagram API Platform. It used to feature basically all possible interactions that can be done with an account, such as accessing all information of public accounts, accessing private content if logged, automate posting for users, etc. On January 30th 2018 it was announced that the API was going to be closed and a full deprecation calendar was shared.[33] At the same, the new Instagram Graph API[34] was released in order to give support only to business accounts that can prove that gathering data and statistics is fundamental for them, otherwise the API is now completely closed to new requests.

3.2.3 Web Scraping

Despite the previous section was one step further into difficulty, it still referred to structured data since, once the researcher learns how to communicate with the database using the protocols, the obtained information is already provided as a table with clear and meaningful variables. We now focus on a more complicated scenario, that of pseudo structured data. The archetypal situation when this happens is looking for information in a webpage where there are plenty of data but there is no download feature nor API for doing it. However, as before, there is a general process to follow:

- First, clearly define what is the information to be downloaded.
- Understand how the webpage is encoding it.
- Select and download it with **R**.
- Transform it into variables and combine them into a data frame.

It is worth mentioning that although a platform might not be specifically interested in easing out process of retrieving structured information to the user, scraping is legal unless explicitly stated otherwise. Every webpage has a `robots.txt` file usually located at the main page, something like www.thewebpage.com/robots.txt and one should always read it before scraping to assess that you are doing a rightful use of the services.

We depict the whole process of web scraping by means of two examples.

---

[31] https://cloud.google.com/maps-platform/?hl=es.

[32] http://geoawesomeness.com/developers-up-in-arms-over-google-maps-api-insane-price-hike/.

[33] https://developers.facebook.com/blog/post/2018/01/30/instagram-graph-api-updates/.

[34] https://developers.facebook.com/docs/instagram-api.

3.2.3.1 TheSportsDB

The first example goes back to TheSportsDB. Usually, there would be no point on scraping a web that has an API, as that method is always preferred, but in this case it serves as a nice example. We first check https://www.thesportsdb.com/robots.txt. It claims

```
User-agent: *
Disallow:

User-agent: AhrefsBot
Disallow: /
```

We start reading the third and fourth line. It means that the robot AhrefsBot is forbidden to scrape anything from the webpage (the symbol/means everything, thus they are disallowing everything). The first and second line say that for all the remaining bots (represented by the wildcard *) nothing is disallowed. That grants legal access, in particular, to the script we are about to create.

We want to download the information on all NBA games in the 19–20 season from https://www.thesportsdb.com/season.php?l=4387&s=1920.[35] In order to start scraping we will need to tell **R** how to browse the Internet and once in the correct link, what to *pay attention to*, so the first step is learning how a webpage is internally referring to each of the elements displayed on screen. Most webs are encoded in HyperText Markup Language (HTML in what follows), a file format specifically designed for the Internet that implements classes and labels so that every text or element in the screen has an identifier. Some browsers, such as Firefox, Chrome, or Opera, allow to find these identifiers. The example continues assuming that Chrome is used, but it can be done very similarly with the other ones.

Surf to https://www.thesportsdb.com/season.php?l=4387&s=1920 and click with the secondary button on the result of the first match and select *Inspect*.[36] A menu should appear in the right-hand side of the screen (see Fig. 3.1), where the code for the whole webpage is displayed. Hovering over the code will highlight the corresponding element in the screen, making it really easy to identify different pieces of information by just scrolling through the code.

All elements in `.html` start and end by a tag. A starting tag always looks like `<tag>`, whereas an ending one is `</tag>`.[37] The starting part of the tag might also

---

[35]TheSportsDB is a quite stable web but with time the tags might change. Should this happen, the reader can download the snapshot of the web as it was when the book was written from our data repository at https://github.com/DataAR/Data-Analysis-in-R/tree/master/webs. Then replace https://www.thesportsdb.com/season.php?l=4387&s=1920 by the route to the downloaded `.html` file in your computer.

[36]Any other match can also be clicked to obtain the equivalent information, but with a different part of the code highlighted.

[37]It is not intended here to learn `.html`, just being able to extract information using tags, for more information on tags check https://www.w3schools.com/html/html_elements.asp.

Fig. 3.1 Inspection menu of an HTML image from TheSportsDB

contain attributes, `<tag attribute=value>`. For our purposes we are only interested in the attribute `class` that works as an identifier of information.

Going back to the example about NBA, take a look at the line of code `<td style="text-align:center; width:10%">96 - 109</td>` highlighted in Fig. 3.1. That tag identifies the result of the first game. Inside the tag, some text describes information about the alignment and some other text is just what it is displayed on the screen. It can always be assumed that contents inside of a certain tag will have the same structure. In other words, all the sets that contain the information about the results are embedded in a `<td>...</td>` tag inside of which the same structure is preserved. This is a key point, since in order to retrieve all the information at once we will refer to the tag `"td"` and **R** will automatically pick the contents of all tags matching it.

Having identified what parts of the code have the information that we are looking for, the next step is to tell **R** how to open a webpage and what to look for in its code. More specifically, for every variable to retrieve, a three-step process is performed. Firstly, we call the webpage and store it in a variable. Then we specify what part of the page is to be extracted, using for that the tags as described above. Finally, the information is converted to text. Type in the console

```
library(rvest)
```

which loads the `rvest` package, specifically designed for interactions with webpages. The command `read_html()` is used to open the webpage from **R**. The

URL is not loaded graphically so it is not displayed as if it was loaded from a standard internet browser. Conversely, when using the command

```
nbagames <- read_html(
  "https://www.thesportsdb.com/season.php?l=4387&s=1920")
```

the whole code of the webpage is saved into a variable. Indeed, by typing nbagames, we get

```
{html_document}
<html>
[1] <head>\n<meta http-equiv="Content-Type" content="text/h ...
[2] <body class="homepage">\n\n\r\n<header id="header"><nav ...
```

The bit of information that we are interested in can be extracted by using the command html_nodes(). This function needs two arguments, firstly the .html to explore, secondly where to look inside the page. The second argument is a string of text that contains the tag to look for. A tag such as <td> is introduced as a string, simply "td". Therefore, writing

```
games <- html_nodes(nbagames,"td")
```

yields

```
{xml_nodeset (7731)}
[1] <td><br></td>\n
[2] <td style="text-align:left; width:20%">04 Oct 2019</td>\n
[3] <td></td>\n
...
```

Indeed, the command imports all the td tags, but the whole chunk of .html code is stored in a xml_nodeset, a class specifically designed for interaction with .html codes. In order to transform it into a character variable that we can easily process, just add the command html_text() to the previous code as follows:

```
games <- html_text(html_nodes(nbagames, "td"))
```

Now, the variable games is

```
[1]  ""                        "04 Oct 2019"      ""
[4]  "\n\t\t\t\n\t\t\tr00 "    "Los Angeles"      "96 - 109"
[7]  " Houston Rockets"        "05 Oct 2019"      ""
[10] "\n\t\t\t\n\t\t\tr00 "     "Indiana Pacers"   "130 - 106"
[13] " Sacramento Kings"       "06 Oct 2019"      ""
[16] "\n\t\t\t\n\t\t\tr00 "     "Boston Celtics"   "107 - 106"
[19] " Charlotte Hornets"      "06 Oct 2019"      ""
[22] "\n\t\t\t\n\t\t\tr00 "     "Golden State"     "101 - 123"
```

This is a more readable version than the one before, but there is still work to do. There are some empty strings and some containing \n, \t or \tr00[38] that are

---

[38]These strings control spacing in .html. For example, \n corresponds to a new line and \t is a tabulation.

Fig. 3.2 Inspection menu of an HTML image from the Goodreads

disturbing the results. Additionally, we want to split the information into different variables to format it as a table. Also, results are interpreted as characters and not as numbers. But there is nothing else that can be done from the point of view of scraping. The information was singled out in the smallest pieces that the `.html` allowed, so the remaining tuning is left for preprocessing methods (see Sect. 3.3).

3.2.3.2 Goodreads

The second task is downloading the list of best books of the twenty-first century according to Goodreads that can be found on https://www.goodreads.com/list/show/ 7.Best_Books_of_the_21st_Century, together with the information about author and ranking.[39] Before we start we check https://www.goodreads.com/robots.txt. Note that now many domains are disallowed for a generic user (denoted by *) but/is not disallowed (so not everything is forbidden) and neither /list/ is, so it is perfectly legal to scrape the page we want to obtain information from it.

As before, prior to writing any code, we analyze the web by inspecting the `.html` in Chrome. Right click on the title of the first book, and inspect the element, to obtain Fig. 3.2.

---

[39] As before, Goodreads is a very stable web but with time the tags might change. Should this happen, the reader can download the snapshot of the web as it was when the book was written from our repository at https://github.com/DataAR/Data-Analysis-in-R/tree/master/webs. Then replace "https://www.goodreads.com/list/show/7.Best_Books_of_the_21st_Century" by the route to the downloaded .html in your computer.

We see a well-defined structure and detect that the tag span seems to enclose the names of the books. Following what we did in the previous example, the code for retrieving the book title should be

```
goodreadsurl <-
  "https://www.goodreads.com/list/show/7.Best_Books_of_the_21st_
  Century"
goodreads <- read_html(goodreadsurl)
books <- html_text(html_nodes(goodreads, "span"))
```

However, this code produces the following:

```
[1]  "Browse "
[2]  "Listen with Audible"
[3]  ""
[4]  "Sponsored"
[5]  "Community "
[6]  ""
[7]  ""
[8]  ""
[9]  ""
[10] "Profile"
[11] "Groups"
[12] "Groups"
[13] "Friends recommendations"
[14] "Friends recommendations"
[15] "Browse "
[16] "Listen with Audible"
[17] ""
[18] "Sponsored"
[19] "Community "
[20] "Harry Potter and the Deathly Hallows (Harry Potter, #7)"
[21] "by"
```

It is only in line 20 that the title of the first book shows up and before that, there is plenty of information that has nothing to do with what we want to obtain. This is an extremely common situation; all concepts of the same type are in the same tag (all books will be listed in a span tag) but not all tags of one kind have just one kind of information (not all span tag are book titles). The fact that in the previous example this did not happen is a mere coincidence, and a lucky one. The problem is that, of course, when dealing with large amounts of data, the analyst cannot check one by one whether the results are the desired ones or not, it has to be known from the code.

To solve this situation, go back to Fig. 3.2 and note that the span tag is inside an a tag and one of the attributes of this tag reads class="bookTitle". In HTML, classes are a way of categorizing tags so that you can use a single tag for different contents. In order to refer to a class, we just add a dot before the name of the class. Classes and tags can be combined in the same call, so the code

```
goodreads <- read_html(goodreadsurl)
books <- html_text(html_nodes(goodreads, ".bookTitle span"))
```

selects all nodes with the class bookTitle and picks the span tag inside them.
Indeed, the desired result is obtained.

```
[1]  "Harry Potter and the Deathly Hallows (Harry Potter, #7)"
[2]  "The Hunger Games (The Hunger Games, #1)"
[3]  "The Kite Runner"
[4]  "The Book Thief"
[5]  "Harry Potter and the Half-Blood Prince (Harry Potter, #6)"
[6]  "Harry Potter and the Order of the Phoenix (Harry Potter,#5)"
[7]  "The Help"
[8]  "A Thousand Splendid Suns"
[9]  "Life of Pi"
[10] "Catching Fire (The Hunger Games, #2)"
[11] "The Girl with the Dragon Tattoo (Millennium, #1)"
[12] "The Time Traveler's Wife"
[13] "The Fault in Our Stars"
[14] "The Da Vinci Code (Robert Langdon, #2)"
[15] "The Road"
...
```

In order to complete the task and retrieve information about the year and rating as
well, we inspect the HTML code below the tag associated with the first group and
expand some of the nodes to find that there is a class authorName with a span
tag inside it containing the author and a class minirating containing the ratings.
Therefore the final code looks like

```
goodreads <- read_html(goodreadsurl)
book <- html_text(html_nodes(goodreads, ".bookTitle span"))
author <- html_text(html_nodes(goodreads, ".authorName span"))
rating <- html_text(html_nodes(goodreads, ".minirating"))
topbooks <- data.frame(book, author, rating)
```

At this point, the data frame still shows badly formatted information in the variable
rating, such as the words or hyphen, that should not be there in the final
version but that cannot be avoided with the scraping. This will be overcome with
preprocessing in Sect. 3.3.

Even though in the first example there were no classes, the best practice is to use
the class information whenever a tag has it. The problem with referring only to
the tag is too general, remember how using the class span in this second example
yielded different kinds of information, not only about book titles. If the tag is used,
the mixture of information is more severe, since common tags such as <div>,
<a>, <span>, or <p> are all over an .html file, so calling html_nodes(web,
"div a") is extremely unspecific and will not narrow down the query to the
desired information.

As it was the case with APIs, most of the difficulty related to web scraping
lays in the understanding of the provider more than the **R** coding itself. When
using APIs it was all about finding the correct URL, whereas now the focus is on
inspecting the webpage in order to find the right identifiers, tags, or classes. Then,
the implementation is quite systematic. There are tools that help finding identifiers

such as SelectorGadget.[40] Even though such scripts might ease out the standard cases, they will suffer with complicated ones, where a closer inspection of the tag is needed, or where many tags are nested so their usage is advisable only under close supervision.

We end the section presenting two main difficulties that arise when web scraping that were not faced here as they are clearly beyond the scope of this book. However, these are becoming more frequent and should be mentioned to have a good understanding of the state of the art in these techniques.

The first limitation is related to how a webpage is programmed. A web might feature contents that are displayed only after some clicks, without any change in the URL. Also, some services require to be logged to access all information. Some pages are programmed using javascript on top of html, in order to add a layer of enhancements to that web. All these problems cannot be addressed with the rvest library. There are other packages, though, such as rselenium, that can overcome all these situations and completely mimic any browsing a human could do, but not without effort. Its installation is not as easy as the other libraries that have been discussed here, as some other files are needed in the computer, and the commands become more proficient, as well as the new interactions with the web.

The second obstacle is a harder one. As discussed when dealing with APIs, some services are specifically interested in keeping the data for themselves. Those platforms work hard to detect and block any kind of automation, making it impossible to retrieve more than a handful of information. As an example (experienced by the authors), there is the case of Instagram. They are specially concerned about massively revealing who follows whom, as that is a way of obtaining user names, and that could simplify spamming and following users without real connection. Even when sophisticated techniques that feature randomness to avoid pattern detection were used, Instagram was able to detect the automation and block the connection before the list of followers of a single user was scraped. This comes as no surprise because, after all, they completely shut down their API, therefore it would be nonsense if the information could be scraped somehow else.

3.2.4 Exercises

Exercise 3.4 This exercise goes back to the OpenSky Network.

(a) Try to retrieve the states of all airborne planes in the last 30 min. Can it be done unauthenticated? Could the request be pulled if we created an authenticated user?

---

[40]https://selectorgadget.com.

(b) Get the states of all planes flying over Madrid region. For that, follow the example about Switzerland in the documentation and replace latitude and longitude by the ones corresponding to Madrid.
(c) The Airbus A380 is the largest passenger aircraft ever created, a wonder of engineering, science, and craftsmanship. The airline Lufthansa has one with icao24[41] code 3C65A5, and Emirates has another one with icao24 896477. Find them and identify the flights they are doing. Come back in around 1 h. Where are they now? Is the trajectory consistent with the planned flight?

Exercise 3.5 We use TheSportsDB again.

(a) Obtain an exhaustive list of all sports available in the API. Which ones are present in the USA? Which ones in Italy? And in China?
(b) List all the teams of all sports in the USA. Furthermore, list all players in all teams in all leagues in the USA. You will need the control structures learned in Sect. 2.3.2. Create a data.frame with the following columns, player, team, league, sport. How many players are there?
(c) Find the match with highest score (of both teams combined) in the NBA during season 2016–2017. Who was the best scorer in that match?

Exercise 3.6 The World Data Bank API is quite complicated. The package wbstats in **R** eases out its use. Find the official documentation of this package and follow it to learn how to use it.

Exercise 3.7 Scrape the list of best films according to RottenTomatoes. It can be found in https://www.rottentomatoes.com/ and then surfing to the menu movies and DVDs and then top movies. Once the page loads, surf to "View all." The final output should be a data.frame with three columns, one containing the name of the film and year, the second displaying the director, and the third showing the cast of the film.

3.3 Data Preprocessing

The algorithms and methods used for data analysis that will be presented in Chap. 5 are powerful and revealing but as important as knowing and understanding the intricacies of them is being able to prepare the data in such a way that the method will reach its full potential. This is a crucial concept to keep in mind, the mere fact that the information of the study to conduct is filed as one dataset or many, or how the variables are structured might determine what can be done or not. Patterns and conclusions that can be inferred with the adequate structure might be lost, worsening the analysis or even crippling its success.

---

[41] The icao24 is a permanent hexadecimal code that identifies every aircraft.

The process of configuring the information in the most suitable way for the statistical methods to be implemented is known as data preprocessing [5] and, among data analysis tasks, this is a highly time consuming one, yet fundamental.

Data preparation is usually described as a four-step process [5, 8] even though not all of them are always needed.

1. **Cleaning**: It is clear from the previous section that even when the importation has been properly done there might be extra characters that should be removed (as the variable `rating` in the Goodreads example in Sect. 3.2.3) or several pieces of information lay in a single column that hinders the analysis (recall the TheSportsDB example in Sect. 3.2.3) or numeric data stored as characters (as it happened again with the TheSportsDB example in Sect. 3.2.3). All the manipulations which purely refer to changing the format of the variables of a dataset is called cleaning.
2. **Integration**: Information might be coming from different places and, after each of those files has been cleaned, joining several files into a single data frame is needed.
3. **Transformation**: Sometimes extra variables that might be relevant can be created, or aggregation is needed or the dataset can be "reshaped" into a format which is better to certain analysis.
4. **Reduction**: When dealing with huge datasets, if the analysis to conduct is somewhat specific, some variables might be skipped freeing memory and simplifying calculations.

Recall that **R** is a free open-source programming language so it is often the case that the same results can be accomplished in several ways by using different packages [12]. In this section we stick to the `base` and `data.table` packages, as almost every part of preprocessing can be carried out only with those two and, in that case, then they outperform other solutions such as `tidyr`, `dplyr`, `stringr`, or `stringi` in terms of speed [1] and simplicity. The advantage of those packages is only noticeable for very advanced features out of the scope of this book.

All the parts of preprocessing are explored throughout this section, following a first part devoted to data tables, the main tool that will help us perform the four steps, with the purpose of finally having a perfectly ready dataset for data visualization and analysis.

3.3.1   Data Tables

We start off with data tables, an improved version of data frames. Recall from Sect. 2.2.5 that an object of the class `data.frame` is a tabulated structure of variables (not necessarily of the same class) that are sorted in columns. In Sect. 3.1 we explored how to import from and export to a data frame using different functions, and the use of the library `data.table` was highlighted for the fast commands `fread()` and `fwrite()`.

A file imported using `fread()` is not only endowed with the `data.frame` structure. In fact, `data.table` is an enhancement of the class `data.frame` oriented to high performance dataset handling with focus on easy coding and, more importantly, functionality and speed. Not only loading to and saving from a data table is faster compared to data frames, also plenty of operations are swift thanks to two major merits of how the package is programmed. On the one hand, adjacent information in data table columns is stored contiguously in RAM memory, so some *under the hood* operations are skipped and some others optimized [3]. On the other hand, the commands for transformations in the data table are implemented by using fast and efficient algorithms such as the Radix sort[42] [6, 11].

A major convenience is that any `data.table` object is also a `data.frame` one, so all the commands and syntax learned in Sect. 2.2.5 are still valid and coexist with the rest of the commands added by the package. From now on, we will assume that whenever we work with a dataset it will be structured as a `data.table`, so the loading is done with `fread()`. An already existing vector or matrix can be coerced into a data table using the command `as.data.table()` as follows:

```
our.matrix.DT <- as.data.table(our.matrix)
```

Existing data frames or lists are also transformed to data tables with the command `setDT()`[43] without the need to assign to a new object. Data tables do not have row names, so if the original data frame had them, we preserve them by using the argument `keep.rownames` that adds a new column with name `rn`.[44] For example, the code

```
library(data.table)
example <- data.frame(info1 = c(1, 2), info2 = c("a", "b"))
row.names(example) <- c("line1", "line2")
setDT(example, keep.rownames=T)
example
```

```
      rn info1 info2
1: line1     1     a
2: line2     2     b
```

```
class(example)
```

```
[1] "data.table" "data.frame"
```

creates a data frame and assigns row names to it. Then it is embedded with the data table structure and the final result is both a `data.frame` and a `data.table`.

---

[42]https://en.wikipedia.org/wiki/Radix_sort.

[43]They can also be coerced with `as.data.table()` but the command `setDT()` is faster and uses less memory.

[44]Not having row names might be confusing at first but it is really an advantage. Either the row name is just the index which is irrelevant information and there is no need to keep it, or it is meaningful information and then it should be treated as any other variable, with the same *column status*.

A data.table object has the format example.data[i, j, by] where the three arguments correspond, respectively, to row, column, and how to group. As explained in Sect. 2.4, if only two arguments are passed, it will be understood as the first two, unless specified. Data tables admit passing just one argument as opposed to data frames where both row and column are mandatory.

3.3.1.1   Ordering

Reordering the rows of a data frame was studied in Sect. 2.2.5. Although the same code can be used, data tables allow for a slightly simpler syntax and the execution is faster, especially in large tables.

Take, as an example, the dataset swiss from the default package datasets which contains data about French-speaking towns in Switzerland in the nineteenth century.[45] In order to avoid messing with the original dataset we create a copy, set it as a data table, and order them alphabetically

```
DT.swiss <- copy(swiss)
setDT(DT.swiss, keep.rownames=T)
DT.swiss[order(rn)]
```

	rn	Fertility	Agriculture	...	Catholic	Infant.Mortality
1:	Aigle	64.1	62.0	...	8.52	16.5
2:	Aubonne	66.9	67.5	...	2.27	19.1
3:	Avenches	68.9	60.7	...	4.43	22.7
...						
45:	Vevey	58.3	26.8	...	18.46	20.9
46:	Veveyse	87.1	64.5	...	98.61	24.5
47:	Yverdon	65.4	49.5	...	6.10	22.5

In addition to the speed and simplicity of the last line of code, note that the alphabetical arrangement is possible thanks to the fact that the names of the towns are now formatted as a column.

Even more, multi-ordering is possible. When several variables are passed as arguments, the table is ordered by the first one, then rows with equal values for that variable are ordered following the second criterion and so on. If the symbol minus ("−") is used in front of a variable it will be sorted in descending order. The following code sorts the table in ascending order with respect to the Education percentages,[46] and in descending order by Agriculture occupation

```
DT.swiss[order(Education, -Agriculture)]
```

	rn	Fertility	Agriculture	...	Catholic	Infant.Mortality
1:	Oron	72.5	71.2	...	2.40	21.0

---

[45]Type ?swiss in the console for details.

[46]The variable Education represents the percentage of military draftees who got an education beyond primary school.

2:	Herens	77.3	89.7 ...	100.00	18.3
3:	Conthey	75.5	85.9 ...	99.71	15.1
...					
45:	Rive Gauche	42.8	27.7 ...	58.33	19.3
46:	Neuchatel	64.4	17.6 ...	16.92	23.0
47:	V.De Geneve	35.0	1.2 ...	42.34	18.0

3.3.1.2 Subsetting

As it was the case with data frames in Sect. 2.2.5, subsetting in a data table by using the row or column indices is done by just referring to them in brackets. For example, to select the values of the seventh column for rows 2 and 3:

```
DT.swiss[2 : 3, 7]
```

```
   Infant.Mortality
1:             22.2
2:             20.2
```

Column subsetting can also be performed by name as follows:

```
DT.swiss[, "Infant.Mortality"]
```

or by *deselecting* a column with the symbol ! in front of the name

```
DT.swiss[, !"Fertility"]
```

or by any combination of the above carried out by c(), for example,

```
DT.swiss[, !c("Agriculture", "Catholic")]
```

	rn	Fertility	Examination	Education	Infant.Mortality
1:	Courtelary	80.2	15	12	22.2
2:	Delemont	83.1	6	9	22.2
...					
46:	Rive Droite	44.7	16	29	18.2
47:	Rive Gauche	42.8	22	29	19.3

In a set with millions of entries, though, the most common scenario is that we look for a particular piece of information neither the index nor the name of where it lays can be known in advance, so conditional subsetting is necessary. The syntax our.data[condition on certain variables] works perfectly, since it will choose those rows where the condition is satisfied, and no extra arguments are needed in a data table. In the following code, those towns where draftees educated beyond primary school account to exactly 7% are selected

```
DT.swiss[Education == 7]
```

	rn	Fertility	...	Education	Catholic	Infant.Mortality
1:	Moutier	85.8	...	7	33.77	20.3
2:	Porrentruy	76.1	...	7	90.57	26.6
3:	Broye	83.8	...	7	92.85	23.6

```
4:        Gruyere       82.4 ...              7      97.67               21.0
5:        Aubonne       66.9 ...              7       2.27               19.1
6:    Val de Ruz        77.6 ...              7       4.97               20.0
7: ValdeTravers         67.6 ...              7       8.65               19.5
```

The function subset () was explained in Sect. 2.2.5 for conditional subsetting and
the same result can be obtained with that approach, but data tables make it easier in
terms of code (compare the syntax above to the one used in for data frames) and are
clearly faster (with small tables no difference is noticeable, but with larger ones the
change becomes huge).

Queries with more than one condition are also possible. The following code
selects those towns where both the Agriculture and Education variables are
above the average (note how we remove the first column of characters)

```
mean.values <- sapply(DT.swiss[, -1], mean)
DT.swiss[Agriculture > mean.values[2] & Education >
          mean.values[4]]
```

```
       rn Fertility Agriculture Examination Education ...
1:    Aigle       64.1         62.0          21         12 ...
2: Avenches       68.9         60.7          19         12 ...
3:    Nyone       56.6         50.9          22         12 ...
4:     Sion       79.3         63.1          13         13 ...
```

Another advantage of data tables is that different objects can be passed as the
second argument, j, not only indexes. One application of this is subsetting in
combination with certain commands. For example, we can compute the mean of
Education for those towns with Fertility measure below 50, with the code

```
DT.swiss[Fertility < 50, mean(Education)]
```

```
[1] 37
```

The same result can be obtained with some other approaches (some of them can
also be fitted in just one line of code) but again, this one takes advantage of the
speed of data tables. Any other suitable command can be used instead of mean(),
for instance, we can count how many towns have less than 10% of draftees with an
Education beyond primary school by using length

```
DT.swiss[Education < 10, length(Education)]
```

```
[1] 28
```

But what if we wanted to do any of the above calculations for each of the levels of
a certain variable? For example, can we list the mean of the Fertility measure
for each Education percentage?

3.3.1.3   Aggregation

Grouping is possible thanks to the third argument of data tables, by. It should be
used in combination with a command that is passed in j, so in order to get the
average Fertility for each one of the Education values, type:

```
DT.swiss[, mean(Fertility), by=Education]
```

```
    Education        V1
 1:        12 68.04000
 2:         9 71.06667
...
18:        53 35.00000
19:        29 43.75000
```

The results, though, are *in order of appearance*. This could be beneficial in some cases, but if not, the output can be sorted out by using order() (as explained above in the argument i),

```
DT.swiss[order(-Education), mean(Fertility), by=Education]
```

```
    Education        V1
 1:        53 35.00000
 2:        32 64.40000
...
18:         2 73.70000
19:         1 72.50000
```

Note how such a simple task can be useful to draw preliminary conclusions. Here, the aggregated table points to an inverse relation between Education and Fertility although a true analysis should be conducted, as it will be seen in Chap. 5.

When grouping with by, the command length no longer solves the problem of counting. The symbol .N is a special command that can be passed as a j argument of a data table in order to count the number of observations inside groups. The following code counts how many towns are in each level of Education and sorts this in an ascending order

```
DT.swiss[order(Education), .N, by=Education]
```

```
    Education N
 1:         1 1
 2:         2 3
...
18:        32 1
19:        53 1
```

Several queries can be performed at once, for example,

```
DT.swiss[order(Education), .(.N, mean(Fertility),
        mean(Catholic)), by=Education]
```

```
    Education N       V2        V3
 1:         1 1 72.50000  2.40000
 2:         2 3 73.70000 74.63667
...
18:        32 1 64.40000 16.92000
19:        53 1 35.00000 42.34000
```

which shows the number of towns for each Education percentage, and the Fertility measure and Catholic population averages, everything sorted by the variable Education. Finally, arguments such as logical expressions can be passed to by. In that case, the veracity of each logical expression is checked and compared to each case of the other statements. In order to keep exploring the hinted inverse relation of Education and Fertility, we check the towns that are above or below 15% and 60, respectively, for each category.

```
DT.swiss[, .N, .(Education < 15, Fertility > 60)]
```

```
   Education Fertility  N
1:      TRUE      TRUE 37
2:     FALSE      TRUE  2
3:     FALSE     FALSE  6
4:      TRUE     FALSE  2
```

In 43 occasions one variable was above the threshold when the other one was below, while for only 4 cases both were either high or low at the same time. As above, this table strongly suggests an inverse relation of the two variables. Nonetheless, it is just an exploratory analysis that should be used to develop our intuition on the dataset but no formal conclusions should be extracted from this kind of tasks.[47]

3.3.1.4 Keying

Subsetting was first introduced back in Sect. 2.2.5 and then improved in terms of speed, flexibility, and possibilities with data tables. Now, we explore *keys*, yet a faster way of subsetting. When a key is set to certain variable in a data table, the whole table is physically reordered in the RAM memory and the order assigned to the rows is stored internally.

To set a key, the command setkey(data.table, key) is used. For example, the code

```
setkey(DT.swiss, Education)
```

sets the variable Education as the key in DT.swiss. No change seems to have happened, but the data table has been reordered by Education:

```
DT.swiss
```

	rn	Fertility	...	Education	Catholic	Infant.Mortality
1:	Oron	72.5	...	1	2.40	21.0
2:	Echallens	68.3	...	2	24.20	21.2
3:	Conthey	75.5	...	2	99.71	15.1
...						
45:	Rive Gauche	42.8	...	29	58.33	19.3

---

[47]In this case, the thresholds for the variables have been chosen arbitrarily, so they might be misleading. Previous knowledge of the analyst about the subject should rule this kind of choices.

```
46:     Neuchatel      64.4 ...      32    16.92              23.0
47: V. De Geneve       35.0 ...      53    42.34              18.0
```

Now, subsetting by rows with lookups regarding the keyed variable is possible just by passing in i the argument to match inside the operator .(). For example, in order to select the towns with Education equaling 3%, use

```
DT.swiss[.(3)]
```

```
          rn Fertility ... Education Catholic Infant.Mortality
1:       Moudon     65.0 ...         3     4.52             22.4
2: Paysd'enhaut     72.0 ...         3     2.56             18.0
3:      Monthey     79.4 ...         3    98.22             20.2
4:       Sierre     92.2 ...         3    99.46             16.3
```

More than one value can be passed using the standard commands for combination, such as c().

```
DT.swiss[.(c(3, 5))]
```

```
          rn ...Examination Education Catholic Infant.Mortality
1:       Moudon ...       14         3     4.52             22.4
2:Paysd'enhaut ...        6         3     2.56             18.0
3:      Monthey ...        7         3    98.22             20.2
4:       Sierre ...        3         3    99.46             16.3
5:Franches-Mnt ...        5         5    93.40             20.2
6:      Cossonay ...      22         5     2.82             18.7
```

The key set for the argument i can be perfectly combined in a natural way with j and by. Try the following possibilities:

```
DT.swiss[.(1 : 3), !c("Examination", "Infant.Mortality")]
```

```
          rn Fertility Agriculture Education Catholic
1:         Oron      72.5        71.2         1     2.40
2:    Echallens      68.3        72.6         2    24.20
3:      Conthey      75.5        85.9         2    99.71
4:       Herens      77.3        89.7         2   100.00
5:       Moudon      65.0        55.1         3     4.52
6: Paysd'enhaut      72.0        63.5         3     2.56
7:      Monthey      79.4        64.9         3    98.22
8:       Sierre      92.2        84.6         3    99.46
```

```
DT.swiss[.(5 : 8), mean(Fertility), by=Education]
```

```
    Education        V1
1:          5 77.10000
2:          6 71.07500
3:          7 77.17143
4:          8 75.92500
```

that produce, respectively, a table displaying towns with Education percentage equal to 1, 2, or 3 and the variables Examination and Infant.Mortality have been deleted, and another table showing the average Fertility measure of towns with Education percentage being between 5 and 8.

3.3.1.5 Updating by Reference

So far, a variety of tools for subsetting, grouping data, or filtering have been studied but a very basic functionality has been skipped: modifying the table itself. As before, the semantics from data frames is inherited and can be used for these purposes, but the recommended practice is to edit by reference: a way of modifying a table that takes full advantage of the speed of data tables by directly writing in the right places, without unnecessary copies of parts of the table. Edition by reference is carried out with the operator : = preceded by the name of the column(s) to edit and followed by the corresponding value(s) to assign. Therefore, creating two new columns with two sets of numbers is done by

```
DT.swiss[, c("new.col.1", "new.col.2"):=list(1 : 47, 51 : 97)]
DT.swiss
```

		rn	...	Catholic	Infant.Mortality	new.col.1	new.col.2
1:	Oron	...		2.40	21.0	1	51
2:	Echallens	...		24.20	21.2	2	52
...							
46:	Neuchatel	...		16.92	23.0	46	96
47:V.	De Geneve	...		42.34	18.0	47	97

Modifying its values is performed as follows:

```
DT.swiss[, c("new.col.1", "new.col.2")
           :=list(101 : 147, 151 : 197)]
```

and deleting existing ones is as easy as

```
DT.swiss[, c("new.col.1", "new.col.2"):=list(NULL, NULL)]
```

Note that the structure is fully consistent, creating the columns when needed and deleting them if the value NULL[48] is assigned.

3.3.2 Merging

A major task in data preprocessing is joining different tables which may have matching fields. Recall from Sect. 2.2 that if two sets with the same variables want to be merged we can just use the command rbind().

```
dataset.1 <- data.table(city=c("Large", "Medium"),
                        population=c(1000000, 250000),
                        km2=c(20, 7))
dataset.2 <- data.table(city=c("Small"),
population=c(50000), km2=c(1))
dataset.final <- rbind(dataset.1, dataset.2)
dataset.final
```

---

[48]Recall from Sect. 2.2.1 that NULL stands for the empty object.

```
       city population km2
1:  Large   1000000  20
2: Medium    250000   7
3:  Small     50000   1
```

Row binding requires that the amount of columns is the same and also that the respective columns in each dataset are of the same kind. Datasets with the same amount of observations in rows can also be merged and, for it, the command cbind() is used as in the following example:

```
dataset.1 <- data.table(city=c("Large", "Medium", "Small"),
                        population=c(1000000, 250000, 50000))
dataset.2 <- data.table(km2=c(20, 7, 1))
dataset.final <- cbind(dataset.1, dataset.2)
dataset.final
```

```
       city population km2
1:  Large   1000000  20
2: Medium    250000   7
3:  Small     50000   1
```

Far more interesting are joins, situations where the tables share some columns or rows, or parts of them [13]. As an example, we create the following tables:

```
dataset.1 <- data.table(city=c("city.1", "city.2", "city.3",
                              "city.4", "city.5", "city.6"),
                        population=c(10000, 20000, 100000, 5000,
                                   30000, 65000),
                        km2=c(1, 0.5, 0.9, 2, 1.2, 3))
dataset.2 <- data.table(city=c("city.1", "city.2", "city.3",
                              "city.7"),
                        airport=c(FALSE, FALSE, TRUE, TRUE))
dataset.1
```

```
       city population km2
1: city.1     10000 1.0
2: city.2     20000 0.5
3: city.3    100000 0.9
4: city.4      5000 2.0
5: city.5     30000 1.2
6: city.6     65000 3.0
```

```
dataset.2
```

```
       city airport
1: city.1   FALSE
2: city.2   FALSE
3: city.3    TRUE
4: city.7    TRUE
```

and consider the different possible ways of merging them with the use of `merge()`.
By simply using

```
merge(dataset.1, dataset.2)
```

```
         city population km2 airport
1:  city.1          1e+04 1.0    FALSE
2:  city.2          2e+04 0.5    FALSE
3:  city.3          1e+05 0.9     TRUE
```

we obtain a table with all the columns, but only the rows that have values for all
columns, the rest being skipped. This is usually called *inner join*.

For a *full join* where all rows are included, filled with NAs when necessary,[49] we
add the argument `all=TRUE`:

```
merge(dataset.1, dataset.2, all = TRUE)
```

```
       city population km2 airport
1:  city.1      10000 1.0    FALSE
2:  city.2      20000 0.5    FALSE
3:  city.3     100000 0.9     TRUE
4:  city.4       5000 2.0       NA
5:  city.5      30000 1.2       NA
6:  city.6      65000 3.0       NA
7:  city.7         NA  NA     TRUE
```

Finally, we might be interested in keeping all rows from the first dataset and only
those adding the matching columns from the second one, or the opposite, preserving
only those in the second table. Those are frequently referred to as *left join* and *right
join* and are obtained with the help of arguments `all.x = TRUE` and `all.y =
FALSE`

```
merge(dataset.1, dataset.2, all.x = TRUE)
```

```
       city population km2 airport
1:  city.1      10000 1.0    FALSE
2:  city.2      20000 0.5    FALSE
3:  city.3     100000 0.9     TRUE
4:  city.4       5000 2.0       NA
5:  city.5      30000 1.2       NA
6:  city.6      65000 3.0       NA
```

```
merge(dataset.1, dataset.2, all.y = TRUE)
```

```
       city population km2 airport
1:  city.1      1e+04 1.0    FALSE
2:  city.2      2e+04 0.5    FALSE
3:  city.3      1e+05 0.9     TRUE
4:  city.7         NA  NA     TRUE
```

---

[49]Recall from Sect. 2.2.1 the meaning of the NA object, reserving a place in tables which stores
non available entries.

In all cases above, if we reverted table ordering when calling the command, for example, `merge(dataset.2, dataset.1)`, the order of the columns would also be swapped but the resulting table would feature the same information in each case, just in a different order.

3.3.3   *Practical Examples*

We now apply the extensive overview on preprocessing that we just carried out in this section to the datasets obtained in Sect. 3.1. Recall that the whole philosophy behind this is to transform the dataset so that it is best suitable for further statistical analysis and visualization. Not only we use the commands from the previous sections but we also introduce some new ones which are specific for data cleaning such as `na.omit()` or `duplicated()` or for text transformation, such as `substr()`, `tstrplit()`, or `grepl()`. Do not miss them, as they are fundamental in any real database where data has not been previously processed for us.

3.3.3.1   Goodreads Dataset Preprocessing

We start with the Goodreads dataset as it is a really easy example inside the first step in preprocessing: cleaning the database. Remember that the code

```
library(rvest)
library(data.table)
goodreads <- read_html(goodreadsurl)
book <- html_text(html_nodes(goodreads, ".bookTitle span"))
author <- html_text(html_nodes(goodreads, ".authorName span"))
rating <- html_text(html_nodes(goodreads, ".minirating"))
topbooks <- data.frame(book, author, rating)
```

saves a table with the title, author, and rating of the top books in the twenty-first century according to Goodreads.[50] The rating column, though, contains extra characters from the automated scrape that should be cleaned, otherwise no calculation can be performed with this variable.

Selecting subsets of a string by position of the characters can be easily performed in **R** with the command `substr(variable, start, stop)` where `start` and `stop` are, respectively, the first and last character of the `variable` to keep. In the Goodreads case, just type

```
topbooks$rating <- as.numeric(substr(topbooks$rating,1,4))
```

---

[50]The ranking is open to user ratings and the displayed table might change with time.

3.3.3.2 TheSportsDB NBA Dataset Preprocessing

The NBA dataset extracted from TheSportsDB is more intricate and the preprocessing to conduct is a bit more elaborated. Remember that the code

```
library(rvest)
library(data.table)
nbagames <- read_html(
  "https://www.thesportsdb.com/season.php?l=4387&s=1920")
games <- html_text(html_nodes(nbagames, "td"))
```

yields a vector containing the information about NBA matches but there were many unwanted entries. We can start by getting read of all the empty entries. Simply, check which values of the vector games are an empty string and negate this selection with the exclamation point:

```
games <- games[!games==""]
```

Now we want to remove all entries containing \n but the problem here is that now not all entries are the same, they have different values (for example, some entries finish with the string tr00 while some others with tr01) and we do not want to look individually at each of them. In order to bypass this situation we use grepl(text-to-find, where-to-find-it).

```
games <- games[!grepl("\n",games)]
```

Now, we only have relevant information, but it is stored as a vector. Since the vector repeats always the same structure (date, team1, result, team2) we can transform it into a four columns matrix and, then, into a data table.

```
games1920 <- as.data.table(matrix(games, ncol=4, byrow=T))
colnames(games1920) <- c("Date","TeamA","Result","TeamB")
```

Finally, no statistics can be performed from the result column, as it features two numbers. Ideally, we should have two columns, so we can work with the figures. Since the information does not always have the same length (sometimes two digits, some other times three) we want to split the information whenever there is a space. For that, we use the command tstrsplit(variable, separator, keep) from the library data.table. It is recommended to first explore how the column will be split, just by choosing the space as a separator:

```
games1920[, tstrsplit(Result, " ")]
```

```
      V1 V2   V3
1:    96  -  109
2:   130  -  106
3:   107  -  106
4:   101  -  123
5:    91  -  115
```

so that we can pre-specify which columns are to be kept and their names, and remove the `Result` column, so the final code looks like

```
games1920[, c("PointsA", "PointsB"):=tstrsplit(Result, " ",
        keep=c(1,3))]
games1920$Result <- NULL
```

```
          Date             TeamA               TeamB PointsA PointsB
1:  04 Oct 2019     Los Angeles  Houston Rockets       96 109
2:  05 Oct 2019  Indiana Pacers  Sacramento Kings     130 106
3:  06 Oct 2019  Boston Celtics Charlotte Hornets     107 106
4:  06 Oct 2019     Golden State     Los Angeles      101 123
5:  08 Oct 2019 Detroit Pistons    Orlando Magic       91 115
   ---
1284:15 Apr 2020         San Antonio     New Orleans      <NA>
1285:15 Apr 2020 Washington Wizards   Indiana Pacers      <NA>
1286:16 Apr 2020       Phoenix Suns     Los Angeles       <NA>
1287:16 Apr 2020     Portland Trail     Los Angeles       <NA>
1288:16 Apr 2020   Sacramento Kings    Golden State       <NA>
```

Of course, the last games have missing information, as they have not occurred as of the writing of this book.

3.3.3.3 OpenSky Dataset Preprocessing

A lot has been said throughout Sect. 3.3.1 concerning processing speed but the `swiss` dataset used as example was small enough to make the remarks on speed noticeable. The choice was such, in order to depict the new concepts in an understandable and intuitive dataset. We now dive in an example with a larger data table, where the differences in speed can already be appreciated. For the sake of simplicity we will only discuss the data table approach, but the reader is encouraged to try the same preprocessing with the data frame semantics.

We will work with the `flights.csv` file that can be found in the link https://github.com/DataAR/Data-Analysis-in-R/blob/master/datasets/flights.csv, a tabulated file containing snapshots of all flights from the OpenSky API taken every 15 min from January the 21st 2019 at 00:00 to January the 30th 2019 at 00:00. It is a ~500 MB dataset of 4,458,484 rows and 18 columns. We load it and check its size.

```
library(data.table)
flights <- fread("flights.csv")
object.size(flights)
```

```
512277600 bytes
```

This means that around 500 MB of the RAM memory of your computer are now being used to keep this table open and ready. First things first, we rename the variables so as to make them meaningful (check OpenSky's API documentation for details)

```
names(flights) <- c("icao24", "callsign", "country",
             "last.update","last.contact", "longitude",
```

```
"latitude", "altitude.baro", "ground",
"velocity", "track", "vertical.rate",
"sensors", "altitude.geo", "squawk",
"spi", "position.source", "time")
```

Depending on the study to carry out, some variables might not be needed. In this example, `sensors`, `squawk`, `spi`, and `position.source` are too technical for our purposes and `altitude.geo` and `last.contact` can be considered redundant since we also have the variable `time`,[51] therefore we can run

```
flights[, c("last.contact", "last.update", "sensors",
            "altitude.geo", "squawk", "spi",
            "position.source"):=NULL]
```

Also, the `ground` variable is redundant with `altitude.baro`. Or at least it would be if the latter was completely accurate, but plenty of NAs are found, for example, (this can be seen by counting them with `sum(is.na(df$altitude.geo))` and some positive altitude values are associated with `ground=TRUE`, which does not make sense. We can correct this with the help of `ifelse()` and then get rid of the redundant variable

```
flights[, altitude.baro:=ifelse(ground == TRUE, 0,
        altitude.baro)]
flights[, ground:=NULL]
```

Note the blazing speed when using the semantics of data tables; more than 4 millions of checks were performed in less than a second. Now the dataset is around 370 MB. For any personal computer, reducing from a 500 MB database to a 370 MB one is not much of a milestone, but a shrinkage of 25% might be of great help in larger databases.

The variable `last.update` is still in a counter intuitive format, but it is a good one, nonetheless. Going back to the API documentation, it is explained that the variable is displayed in "Unix timestamp," that is seconds since January 1st, 1970. The column is perfectly suitable for calculations but if we are interested in obtaining a more intuitive one, we can shift the origin to January 21st, 2019 at 00:00,[52] that is, the first timestamp of the data base.

```
flights[, time:=time - 1548028800]
```

Some transformations of variables and reduction of columns have been made but still some preprocessing is needed. What remains is usually the most difficult part in preprocessing, not because of the difficulty itself, which is similar, but because it is usually hard to realize of what needs to be done. A very frequent problem is the missing values which are imported as NA. When one of the variables of an observation is NA we have to decide what to do. If there is any clue of what the original value was, we can solve it manually. Missing values can also be replaced

---

[51] The three variables are not the same, and in some cases they have very different values, but we just keep the last one for the sake of simplicity.
[52] This date in Unix timestamp is 1548028800.

by the mean or median, to be introduced with detail in Sect. 5.1. Nonetheless any choice will have an impact on the final analysis. Possibly the most neutral approach, and a very common one, is to consider any observation (any row) with a NA defective and thus eliminate it completely from the study as it cannot be trusted. This can be easily performed with na.omit()

```
flights <- na.omit(flights)
```

Duplicated rows are also a major problem that can be solved by means of the command duplicated(). In the case of the flights data base, the problem is even trickier as duplication occurs in some columns only. It was the case during the data gathering that the transponder of certain planes was stuck transmitting the same latitude and longitude for many different times, which makes no sense. Searching for duplicates in a whole row will give none, as the time was correctly measured, so the duplicated() command should be used only in the adequate columns. Once we identify the rows with repeated information we take the rest with the command !

```
duplicated.rows <- duplicated(flights[, 1 : 5])
flights <- flights[!duplicated.rows]
```

Further inspection of the database reveals that some callsigns are erroneous or used for testing purposes. We delete all those options with

```
flights <- flights[!callsign == ""]
flights <- flights[!callsign == "0000000"]
flights <- flights[!callsign == "00000000"]
flights <- flights[!grepl("TEST", callsign)]
```

In the first three cases we locate the rows matching the unwanted values and disregard them. The fourth line uses grepl(), the previously mentioned command for partial matching of text. In this case, strings such as TEST1234, 12TEST12 will be selected and thus removed from the database.

The database is now a table of 3,822,909 rows and 10 columns, and is ready to be analyzed.[53]

3.3.3.4 Additional Remarks on Speed

R is considered to be a rather slow language and it is true for most interactions but when using the library data.table it becomes blazing fast because of the way it stores and uses the information in RAM memory. When using this package properly and if no commands from other libraries are incorporated (that might act as a bottle neck slowing down the whole process) then **R** is actually really fast, the fastest among mainstream solutions for data analysis [7]. Only SQL[54] beats **R**'s data

---

[53]In fact, more preprocessing can be conceived, see the Exercises, but this is enough to confidently start the analysis.

[54]SQL is a language designed for database handling, but not for data analysis. Given a database, it constructs an ad hoc predefined hierarchy for it. Since this structure is created specifically, the

tables in some scenarios for accessing or grouping information [9], but SQL lacks of features for preprocessing or data analysis, it is more of a tool for accessing and grouping data.

3.3.4 Exercises

Exercise 3.8 This exercise goes back to the dataset retrieved from RottenTomatoes in Exercise 3.7. Do the right preprocessing:

(a) Transform the data table so that the first column is split into name and year and the third column is divided into as many columns as members of the cast.
(b) Rename the columns with relevant labels.
(c) Clean all remaining unwanted characters.
(d) Coerce the columns to numeric when possible.

Exercise 3.9 This exercise goes back to the dataset retrieved from Goodreads in Sect. 3.3.3. Do the right preprocessing:

(a) Transform the data table so that the rating column is splitted into the average rating and the amount of reviews. Many undesired columns will appear. Delete them as well as the original rating column.
(b) Get rid of the commas in the number of reviews and then coerce both new columns into numbers.

Exercise 3.10 Recall how to import the Google stock value data into a data frame from the Quandl API with

```
fread(URLencode("https://www.quandl.com/api/v3/datasets/
                WIKI/GOOG.csv?start_date=2015-01-01"))
```

and practice the following preprocessing:

(a) Look for NAs or duplicated rows. Delete them if any.
(b) Check that columns Ex-Dividend and Split Ratio feature a single value. Remove those to columns as well as all those with adjusted values.
(c) Transform the data table so that the first column is split into year, month, and day, naming them with relevant names.
(d) Coerce the new columns to numeric.

---

speed for accessing data or editing single fields is huge. However, the rigid structure suffers if big changes are made to the database or large-scale edition is the goal, two major drawbacks in modern data analysis.

Exercise 3.11 We already cleaned the `flights` database but further preprocessing can be performed. This is an advanced exercise.

(a) Even though we removed all duplicated rows, two rows might contain duplicated information in latitude, longitude, and altitude but different timestamps so they are not removed. They should not be considered for the study, as a plane cannot stop in the air, which implies that a technical issue happened in relation to the transponder or similar. Remove all rows for which data in columns 1–5 are identical.
(b) Some callsigns are under 4 character, a non-valid format. Remove those rows (you will need to `apply()` the function `nchar()`).
(c) Identify the callsigns that appear only once. We can consider those as anomalies, since no flight is airborne for less than 30 min in commercial aviation. Remove those lines (consider keying).

References

1. Appsilon Data Science—R language (r-bloggers user). Fast data lookups in R: dplyr vs data.table. https://www.r-bloggers.com/fast-data-lookups-in-r-dplyr-vs-data-table/, 2019. [Online, accessed 2020-02-29].
2. Domo, Inc. Data Never Sleeps 6.0. https://www.domo.com/learn/data-never-sleeps-6, 2018. [Online, accessed 2020-02-29].
3. Dowle, M., Srinivasan, A., et. al. FAQ about data.table. https://cran.r-project.org/web/packages/data.table/vignettes/datatable-faq.html, 2018. [Online, accessed 2020-02-29].
4. EMC Education Services. *Data Science and Big Data Analytics: Discovering, Analyzing, Visualizing and Presenting Data*. Wiley, New York, USA, 2015.
5. Garca, S., Luengo, J. and Herrera, F. *Data Preprocessing in Data Mining*. Springer Publishing Company, Incorporated, New York, USA, 2014.
6. Gibbons, A. and Rytter, W. *Efficient Parallel Algorithms*. Cambridge University Press, Cambridge, USA, 1988.
7. h2oai on GitHub. Database-like ops benchmark. https://github.com/h2oai/db-benchmark/, 2018. [Online, accessed 2020-02-29].
8. Pyle, D. *Data Preparation for Data Mining*. Morgan Kaufmann Publishers Inc., California, USA, 1999.
9. rikunert.com. SQL versus R - who is faster? http://rikunert.com/sql_r_benchmarking, 2017. [Online, accessed 2020-02-29].
10. Schäfer, M., Strohmeier, M., Lenders, V., Martinovic, I. and Wilhelm, M. Bringing up Open-Sky: A large-scale ADS-B sensor network for research. *Proceedings of the 13th IEEE/ACM International Symposium on Information Processing in Sensor Networks (IPSN)*, pages 83–94, 2014.
11. Sedgewick, R. *Algorithms in C++—Parts 1–4: Fundamentals, Data Structures, Sorting, Searching*. Addison Wesley Professional, Massachusetts, USA, 1999.
12. Wickham, H. *R Packages: Organize, Test, Document, and Share Your Code*. O'Reilly Media, California, USA, 2015.
13. Wickham, H. and Grolemund, G. *R for Data Science: Import, Tidy, Transform, Visualize, and Model Data*. O'Reilly Media, Inc., California, USA, 2017.

Chapter 4
Visualization

Data visualization is a fundamental skill in data science. It can be used as the finest tool for communication of results and also serves for exploratory analysis. A clear and efficient communication of our data is fundamental if we want to transmit our reasoning and convince others of the conclusions of a project. After the data preprocessing is finished, an initial visualization is always recommended since it will shed light about where to look at in our analysis, hinting what kind of methods should be performed.

The researcher can use graphics to represent the data content in a way that allows deeper interpretation and analysis. When this is correctly done, complex data becomes more accessible and unappreciated features often show up, suggesting new ideas and research directions. Visualizing our dataset under different representations while new calculations are performed is usually a most necessary step to keep developing our analysis correctly.

Initial exploratory analysis comes with natural observation and can be substantially helped with visual plots. This is why visualization techniques precede statistical analysis in Chap. 5. However, high quality data analysis has to be complemented with visualization tools to be completely explained and communicated to a broader audience and so, once the statistical study is complete, the reader should come back to this chapter. In that sense, visualization is a complementary tool that accompanies the researcher as soon as the preprocessing is finished, first for exploration, then as a tool for gaining insight and finally as a fine way of communication.

R is a widely used language for data visualization and features a very robust base plot tool that eases out the understanding through graphics. On top of that, several packages have been developed for highly elaborated graphics. We will cover the basic functions as well as three main libraries for sophisticated plots, namely ggplot2, plotly, and leaflet. The last part of the chapter applies the techniques to the flights database that we scraped and preprocessed in Chap. 3.

© Springer Nature Switzerland AG 2020
A. Zamora Saiz et al., *An Introduction to Data Analysis in R*, Use R!,
https://doi.org/10.1007/978-3-030-48997-7_4

4.1 R Base graphics

The main **R** plotting package is `graphics`, a widely used option for creating fast and easy plots. This package is already installed in our computer and loaded by default as part of the core libraries, so it can be used without previous requirements.

The most important function in this package is `plot()`. This is the main plotting tool in **R** and it can produce graphics for an amazing range of objects, not only the common ones, such as vectors and data frames, but also output objects produced by many different functions, either from **R** base packages or additional libraries with very specific aims.

In RStudio, plots appear on the bottom right window inside the *Plots* tab, which includes several file management options. There are several gadgets in the toolbar of this window that we can use without any script programming. By clicking on `Zoom` we can enlarge the current plot into a bigger separated window, so we can visualize our plots with greater detail. Next to it we have the `Export` option, which can be used to save the plot as an image in different formats (this will be further explained in Sect. 4.1.3). The remaining buttons can be used to `Remove` the current plot or `Clear` all plots in the history of the **R** session. Plots can be called with commands inserted in the console, or executed in a `.R` file. When using the R Markdown `.Rmd` file, the plot view is integrated after the code within the file window.

This section starts with `plot()` in Sect. 4.1.1, where relevant arguments and their usage are explained. Function `plot` allows to plot vectors and matrices, hence to represent series of data and cross graphics of two variables. In Sect. 4.1.2 we present other plotting functions for univariate[1] datasets. Then, Sect. 4.1.3 allows to insert legends and texts in the graphical environment and manipulate the graphical parameters to present information in a more fashionable way, plus how to export a **R** plot for further uses.

4.1.1 The Function plot

The basic plot command in **R** is the robust and versatile function `plot()`. A huge variety of objects, even those from very specific libraries, can be passed as arguments for this function to produce a graph. Customization is rather high for a function with so many possibilities but, of course, graphics created by `plot()` do not reach the beauty of graphics obtained with the libraries in the following sections. As such, it is a fundamental tool for exploratory analysis and for plotting while the study is being carried out, because it allows to graphically represent almost any object we are dealing with.

---

[1]A univariate dataset consists of one variable data, whereas multivariate allows for many variables. More on this will be seen in Chap. 5.

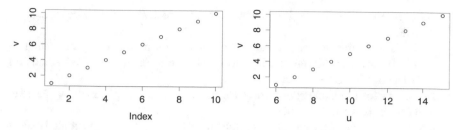

Fig. 4.1 Left: plot of vector v. Right: plot of vector v against vector u

We start off by explaining how to use plot() on objects that we already know from Chap. 2, beginning with vectors. The image on the left of Fig. 4.1 plots a circle dot at the corresponding height for each coordinate of v, whereas the right image uses u as the X-coordinates and places v as the Y-coordinates of plot().

```
v <- 1 : 10
plot(v)
u <- 6 : 15
plot(u, v)
```

This is the basic *scatterplot*, representing two variables in Cartesian coordinates (X vs Y), each pair of values corresponding to the point coordinates. When a single vector is passed to the function, it is interpreted as the vertical coordinates and the corresponding horizontal values are filled with the vector of consecutive numbers 1:n.

Most visualizing functions contain a huge number of arguments leading to many possibilities for plots. General ones, such as colors, labels, and titles are common to most of them,[2] while some are very particular of a specific plotting function. Throughout this book, we do not intend to fully describe all of them, but the most important and illustrative ones. To go beyond that, **R** help can be consulted.

In particular, plot is one of the functions with higher number of arguments in **R**, so we inspect the most fundamental ones to have an overview of the different options.

- x is the vector of points to be plotted. It can also admit a matrix, data frame, or other **R** structures.
- y represents the Y-coordinates, if the object x can act as the abscissa coordinates of the plot.
- pch changes the default plotting dot to other symbols. It uses numbers from 0 to 25 (for example, pch=16 are solid circles) or a character (pch="+"). The default value is pch=1 for an empty circle.

---

[2]Throughout Chap. 4, whenever this happens, we omit repeated arguments and focus only on the particular ones. The reader should understand a similar usage for those arguments appearing in several plotting functions.

- cex rescales the dot size. It is given as a ratio which takes the value cex=1 by default.
- col changes the dot color. As values, it admits names of colors (such as col="blue") or the internal **R** color-coding (for example, col=2 stands for red). Default is col=1 black.
- bg serves for the background color for filled dots, those with pch = 21:25, and default is white.
- main is the character string to be the main title of the plot. Default value is main=NULL (no title).
- xlab, ylab are the names of the X and Y axes. By default they are the names of the plotted vector variables and can be removed by setting xlab="", ylab="".
- xlim, ylim set the limits of the X and Y axes. Requires a vector of two entries indicating the beginning and ending of the plot rectangle. When not specified, **R** calculates a reasonable size for the plot depending on the data.
- type encodes the kind of plot we want; options are "p" for points (default), "l" for lines, "b" for both (points and lines) and "h" for heights or vertical bars.
- lty changes the line type from solid (lty=1 by default) to different dashed types (it goes up to the value 6).
- lwd changes line width, being 1 by default.
- axes is a logical argument that, when set to axes=FALSE it removes axes and box. If we want to remove just one axis, set xaxt, yaxt to xaxt="n," yaxt="n." On the other hand, bty="n" gets rid of the box.

Once plot() is called over an object x, a graph is displayed in the corresponding window and this plot is an active object, so the **R** session considers that further elements can be added until another non graphic command is called. If we want to plot over an active graph, for example, to add a new data series, when calling plot again, use the argument add=TRUE. If we type add=NA, a new graph is created with the previous environment, and if add=FALSE (default), a totally new plot is created.

We will not detail the totality of the arguments above. Some of them are not even direct arguments of the function plot() but they are inherited from other secondary functions called by plot(). However, we will elaborate a bit about some of the options for visual enhancement of the plot() output such as colors, shapes, or sizes.

Color and shape variables accept different values or key names that correspond to a variety of options and styles. Figure 4.2 below shows some of these options. The top part corresponds to the 26 types of point markers available with the flag pch. The middle part shows how color is changed with col. Although hundreds of colors can be used through the RGB[3] system, we focus on the 9 basic colors that are defined

---

[3]RGB stands for Red Green Blue and is a way of defining almost every color based on the proportion of each primary color.

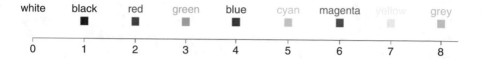

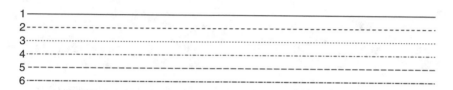

Fig. 4.2 Plot options for dot style (pch), color (col), and line style (lty) with corresponding value and key name

in **R** by name and by numbers (1 corresponds to black, 2 to red, etc.). Resizing the markers is possible by specifying the ratio using cex. While cex=1 is the default size, bigger numbers will increase the size of the dot, and numbers between 0 and 1 will show smaller dots. Negative numbers are not supported as they make no sense in here. The last one of the styling feature displayed in the figure is line type. When representing the evolution of a numerical variable it is usual to connect horizontally adjacent values with lines. First, we need to include the flag type="l" to indicate that values have to be connected with lines. Then, we use lty to choose the line style; solid, dashed, dotted, dotdash, longdash, or twodash. Finally, the line width is specified with lwd as we did with cex, with value 1 as default.

Other plot arguments can be used to configure the *environment* of the output. We can add labels for the X and Y axes and a main title using xlab, ylab, and main, respectively, passing the desired text between quotation marks. In order to set the limits of both axes, xlim and ylim are used. Finally, xaxt="n" and yaxt="n" remove the X and Y axes, respectively, while bty="n" eliminates the box around the graph. Or we can simply remove the three of them with axes=FALSE.

Now that we know the specifics for visual customization, we work again with u and v above and modify the aesthetics in four different ways. The resulting graphs can be seen in Fig. 4.3.

```
plot(u, v, type="l", lty=3, lwd=4, cex=3, col="blue", xaxt="n",
    yaxt="n")
plot(u, v, pch=4, cex=1, col=3, main="Test plot",
    xlab= "Variable X", ylab="Variable Y")
```

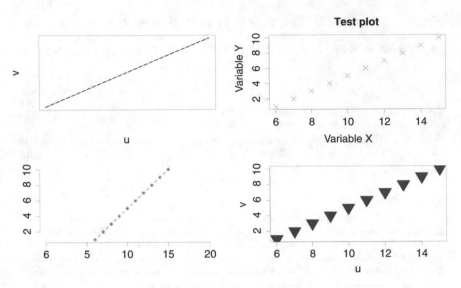

Fig. 4.3 Graphs of vector v against vector u using different arguments of the function plot()

```
plot(u, v, pch=8, type="b", cex=0.8, col=6, xlab="", ylab="",
    bty="n", xlim=c(0, 20))
plot(u, v, pch=25, cex=3, col=4, bg=2)
```

4.1.1.1 Scatterplots

Beyond prettifying arguments, it is utterly important to stress again the robustness of the argument x. It does not only accept coordinates, it is able to work with a huge variety of objects of very different kind ranging from vectors, to data frames or fine structures for multidimensional plotting. This is the main strength of plot() and it is also the reason why it is the most used tool for visualization while the research is being conducted. We now show how plot() deals with more advanced objects.

When the input is a matrix, plot() will use the first two columns as vectors, skipping the rest.[4] Furthermore, it is of special interest the case when a data frame is passed. It creates a multiple image plot where all variables are plotted against all other variables.[5] We now use the iris dataset,[6] a data.frame containing

---

[4]For example, plotting a matrix with u and v as columns yields the right-hand picture in Fig. 4.1.

[5]This can also be obtained with function pairs() from package graphics used on numerical matrices.

[6]The dataset iris is contained in the package datasets included in the **R** core.

measurements of 150 flowers which was introduced by Fisher in [7] and that has
been widely used ever after as an archetypal example. Looking at its structure,

```
str(iris)
```

```
'data.frame':   150 obs. of  5 variables:
 $ Sepal.Length: num  5.1 4.9 4.7 4.6 5 5.4 4.6 5 4.4 4.9 ...
 $ Sepal.Width : num  3.5 3 3.2 3.1 3.6 3.9 3.4 3.4 2.9 3.1 ...
 $ Petal.Length: num  1.4 1.4 1.3 1.5 1.4 1.7 1.4 1.5 1.4 1.5 ...
 $ Petal.Width : num  0.2 0.2 0.2 0.2 0.2 0.4 0.3 0.2 0.2 0.1 ...
 $ Species     : Factor w/ 3 levels "setosa",..: 1 ...
```

five variables can be identified. Four of them are numerical: Sepal.Length,
Sepal.Width, Petal.Length, and Petal.Width, and contain information
about the dimension of flower's petals and sepals; the fifth one is categorical (a
factor) and contains the Species of the flowers: setosa, versicolor, and
virginica. The plots in Fig. 4.4 show the difference when plotting a matrix and
a data frame. When iris is treated as a data frame, graphs relating all variables are
shown at once, whereas after converting it into a matrix, a scatterplot of the first two
columns is displayed. Note that, when plotting a data frame, plots in positions that
are symmetric with respect to the main diagonal contain the same information with
swapped axes.

```
plot(iris)   # generates a multiple image plot
matrix.iris <- as.matrix(iris)   # forces iris as a matrix
plot(matrix.iris)   # scatterplot of the first two columns
```

Thanks to the combined display in the left of Fig. 4.4, relations between some
variables are hinted. For instance, in this particular case, the relationship between
Sepal.Length and Petal.Length (first row, third column) seems to produce
two clearly differentiated groups while the scatterplot between Petal.Length
and Petal.Width (third row, fourth column) points to a relation following
approximately a straight line (with a gap). By just running plot() on a data frame,
relevant information is usually revealed, providing the analyst with ideas about how
to understand the data content and how to continue. This example will lead us to
have separate discussions for petal and sepal lengths, depending on the species, or
to explore the same linear relation for petal lengths and widths in the three species.

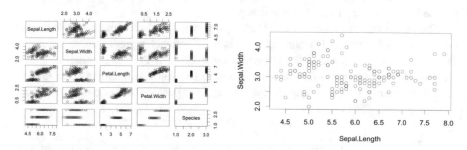

Fig. 4.4 Left: pair scatterplots of data frame iris. Right: scatterplot of the first two columns of
the matrix matrix.iris

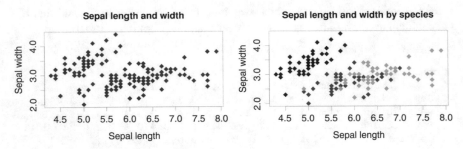

Fig. 4.5 Left: scatterplot of sepal and petal lengths. Right: scatterplot with different colors for species

We now show an important feature of the `col` argument for further customization of the plot. We display again the scatterplot of the sepal length and width where the dots are now blue diamonds (with `col="blue"` and `pch=18`) with a bigger size (using `cex=1.5`). Titles for the whole image and the label axes are also added, so the code is as follows

```
plot(matrix.iris, pch=18, cex=1.5, col="blue",
     main="Sepal length and width", xlab="Sepal length",
     ylab="Sepal width")
```

and the output is shown in the left part of Fig. 4.5. This scatterplot suggests two groups of points, which might come from a different relation between sepal and petal lengths for flowers of distinct species. We will plot those points coming from each specie in a different color, by setting `col=iris$Species`, which forces **R** to pick a different color depending on the value of a third variable,[7] so the code

```
plot(matrix.iris, pch=18, cex=1.5, col=iris$Species,
     main="Sepal length and width by species",
     xlab="Sepal length", ylab="Sepal width")
```

produces the right picture in Fig. 4.5.

It is now obvious that the three species are plotted in different colors and one of them (which can be further confirmed to be `setosa`, the black dots) is essentially different from the other two (`versicolor` and `virginica`, in green and red); setosa flowers have shorter but wider sepals.

The use of variables as customization arguments, such as in this case with the use of `species` for the `col` argument, allows to effectively represent more variables. Here, two numerical variables (`Sepal.Length` and `Petal.Length`) and a categorical one (`Species`) are represented in the same two dimensional plot in an effective and revealing way. A detailed and deeper discussion on the advanced

---

[7]When the argument `col` is filled with the variable `Species`, which is a factor vector with three levels, the first three different colors in the **R** palette are assigned to corresponding observations from each level.

use of customization arguments for introducing more variables in a plot will be presented in Sect. 4.2.

4.1.1.2 Additional Series in a Plot

Sometimes, we want to plot in the same image more than one data series, to understand correlations between values. We will add new information to a given plot, in the form of a series of points or lines. As we mentioned before, after function plot () is called, the plot is an active object in the **R** session and can be modified with several other functions, for example, adding points and lines with points () and lines (), respectively.

This toy example can help to understand how to use points (). It first creates the vector x and y1. Then, vectors y2, y3 and y4 are defined by shifting y1 by 1, 2 and −1 units respectively. Then, it plots x and adds the other vectors with points () which admits the argument type (previously explained for plot ()).

```
x <- 9 : 15 / 2
y1 <- c(2.6, 2.8, 3.2, 3.6, 4, 4.2, 4.3)
y2 <- y1 + 1
y3 <- y1 + 2
y4 <- y1 - 1
plot(x, y1, type="p", lty=1, lwd=2, xlim=c(4, 8), ylim=c(0, 6.5),
     ylab="Offset")
points(x, y2, type="l", lty=1, lwd=2)
points(x, y3, type="b", lty=1, lwd=2)
points(x, y4, type="h", lty=1, lwd=2)
```

The resulting plot can be seen in Fig. 4.6. Note how we previously define the *x* and *y* limits for the plot window, taking into account the values that the other series will reach. Note that, when calling points (), we do not include the argument add=TRUE, it automatically adds a new series to the active plot; the same will apply to lines ().

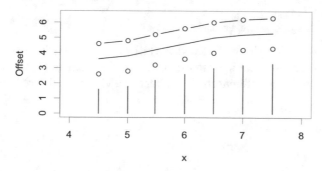

Fig. 4.6 Several time series using function points ()

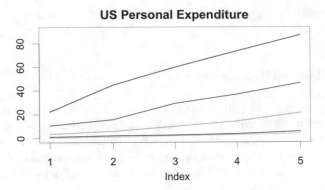

Fig. 4.7 Several time series plot with function `lines()`

A more practical example is the usage of `lines()` when working with time series. We use now the built-in dataset `USPersonalExpenditure`[8] that contains the expenses of the average US citizen in five goods or services at 5 different year times in the period 1940–1960.

`USPersonalExpenditure`

	1940	1945	1950	1955	1960
Food and Tobacco	22.200	44.500	59.60	73.2	86.80
Household Operation	10.500	15.500	29.00	36.5	46.20
Medical and Health	3.530	5.760	9.71	14.0	21.10
Personal Care	1.040	1.980	2.45	3.4	5.40
Private Education	0.341	0.974	1.80	2.6	3.64

This dataset is a matrix object and we can easily represent the evolution of each variable with one call per row and display them in the same image (see Fig. 4.7).

```
plot(USPersonalExpenditure[1, ], type="l", lwd=2,
     ylim=c(0, 90), ylab="", main="US Personal Expenditure")
lines(USPersonalExpenditure[2, ], col=2, lwd=2)
lines(USPersonalExpenditure[3, ], col=3, lwd=2)
lines(USPersonalExpenditure[4, ], col=4, lwd=2)
lines(USPersonalExpenditure[5, ], col=5, lwd=2)
```

The first call (function `plot()`) draws the black line between the specified *y* limits and adds the main title. Calling, after that, the function `lines()`, we add each one of the other series in a different color, completing a vision of all variables in a single plot.

---

[8]The `USPersonalExpenditure` dataset is contained in the `datasets` package.

4.1.1.3 Drawing Mathematical Functions

Although **R** is specially suited for data analysis and statistics, the graphics package also includes a function for drawing custom mathematical functions. With function curve() we can easily evaluate and draw a user defined or **R** built-in function. The code

```
curve(sin, from=0, to=2 * pi, lwd=2, lty=2)
```

plots the function $sin(x)$ from the value 0 to the value 2π (see Fig. 4.8). If limits are not assigned, the plot is drawn between 0 and 1. **R** builds a plot which is sufficiently smooth for the human eye, although it is the result of joining the segments between the discrete evaluation of the function in a series of points. If we want to specify the number of points, we use the argument n. Apart from this, the rest of plot arguments apply as well.

As another example for the use of curve with a non-standard function, suppose that we are studying the equilibrium point of the price of some good which is found as the crossing of its supply and demand curves. Consider that we know the supply and demand prices as functions $s(q)$ and $d(q)$ of the quantity produced q (for example, $s(q) = q^2 + 2$ and $d(q) = 5 + q - q^2$), and implement these functions in **R** with the code

```
supply <- function(x) {x ^ 2 + 2}
demand <- function(x) {5 + x - x ^ 2}
```

We plot these two functions with curve() and recognize the equilibrium price as the intersection of the two curves in Fig. 4.9. Note that we use the argument add=TRUE to overwrite the second curve in the active plot.

```
curve(supply, from=0, to=3, ylim=c(0, 6), lwd=2,
      col="blue", main="Equilibrium price", xlab="Quantity",
      ylab="Price")
curve(demand, add=TRUE, lwd=2, col="red")
```

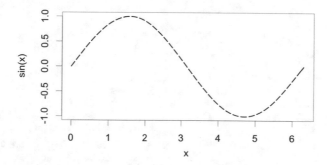

Fig. 4.8 Function $sin(x)$ plotted with curve()

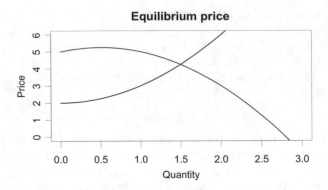

Fig. 4.9 Equilibrium price as the intersection of supply and demand curves using `curve()`

Finally, Fig. 4.2 at the beginning of the section was presented without the code that generated it, as it was too complicated with the knowledge we had back then. We now include the commands needed for generating the three plots. Here we can observe most of the possibilities offered by `plot()` and how to add new series with `points()`. How to add text with `text()` will be explored in Sect. 4.1.3.

```
# First part
plot(0 : 25, rep(0, 26), pch=c(0 : 25), cex=3, yaxt="n", ylab="",
    xlab="", col=c(rep(1, 21), rep(2, 5)), bg=3, bty="n")
axis(side=1, at=seq(1, 25, 1), labels=c(1:25))

#Second part
labels <- c("white", "black", "red", "green", "blue", "cyan",
           "magenta", "yellow", "grey")
plot(0 : 8, rep(0, 9), pch=15, cex=3, yaxt="n", ylab="", xlab="",
    col=0:8, bty="n")
text(0 : 8, rep(0.8, 9), labels, col=c(1, 1 : 8), cex=2)
axis(side=1, at=seq(1, 25, 1), labels=c(1 : 25))

#Third part
labels <- c("6", "5", "4", "3", "2", "1")
A <- matrix(NA, 12, 2)
A[, 2] <- c(6, 6, 5, 5, 4, 4, 3, 3, 2, 2, 1, 1) * 0.3
A[, 1] <- rep(c(0.2, 1), 6)
plot(A[1 : 2, ], type="l", lty=1, lwd=2, ylim=c(0.1, 1.9),
    bty="n", axes=FALSE, ylab="", xlab="")
text(rep(0.2, 6), (1:6)*0.3, labels, cex=2)
points(A[3 : 4, ], type="l", lty=2, lwd=2, ylim=c(0.1, 0.6))
points(A[5 : 6, ], type="l", lty=3, lwd=2, ylim=c(0.1, 0.6))
points(A[7 : 8, ], type="l", lty=4, lwd=2, ylim=c(0.1, 0.6))
points(A[9 : 10, ], type="l", lty=5, lwd=2, ylim=c(0.1, 0.6))
points(A[11 : 12, ], type="l", lty=6, lwd=2, ylim=c(0.1, 0.6))
```

4.1.2 *More Plots for Univariate Series*

Often, datasets we handle are made of single series of values. These are called *univariate series* because they depend on a unique variable. Representing graphically these series can be done in different ways, based on the particular characteristics we want to stress.

Scatterplots above are one way of representing univariate series, but in this section we introduce plots based on the notion of frequency, the number of times each value is repeated in a series of data. Bar plots are basic ways to represent this, by means of a bar of a height proportional to the frequency of each value. Histograms are close relatives, for distributions of continuous values where we are interested in knowing the frequency of a given interval, instead of a discrete number. Pie charts are common plots displaying relative frequencies. Box plots are a bit more evolved graphs comprehending a series of position measures in a distribution, indicating whether values are more or less disperse and/or skewed.

R features several functions for univariate data in the `graphics` package, such as `barplot()`, `dotchart()`, `boxplot()`, `hist()`, and `pie()`, to reproduce bar plots, dot charts, box-and-whisker plots, histograms, and pie charts, respectively.

4.1.2.1 Bar Plot

One of the essential tools for visualizing a series of values is the bar plot. It compares values in a series of data by showing a bar whose height is the given value. The function `barplot` from the package `graphics` takes a numerical vector and plots a series of bars of the corresponding heights.

```
barplot(USPersonalExpenditure[, 5], ylab="Billions of dollars",
        main="US Personal Expenditure in 1960", col="darkgreen")
```

Figure 4.10 shows the different amounts of expenditure in the various goods and services in 1960, where we note that food and tobacco is, by quite a distance, the most common expense. As we can see, many arguments from `plot()` can also be used in this and other graphical functions.

When called on a matrix instead of a vector, `barplot()` produces stacked plots for each column, where the parts of each bar are proportional to each row entries in the matrix (see Fig. 4.11). Stacked bars can be substituted by juxtaposed bars with `beside=TRUE`.

```
barplot(USPersonalExpenditure, legend.text=TRUE,
        args.legend=c(x=2, y=150, cex=0.5))
```

Other arguments of `barplot()` are `width` and `space`, which are used to define the width of each bar and the space between bars. Also, `names.arg` is used to change the names under each bar (by default, vector labels or column labels are

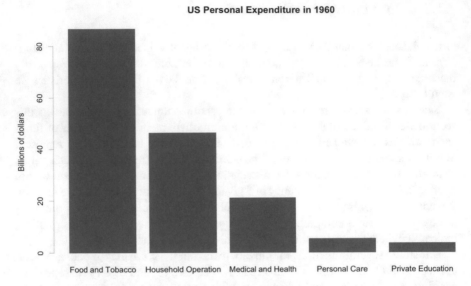

Fig. 4.10 Bar plot of the US Expenditure evolution 1940–1960

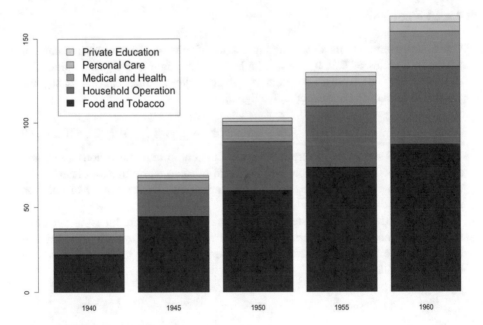

Fig. 4.11 Bar plot of the US Expenditure in 1960

shown). When we produce a bar plot of a matrix, a legend with the contents of each bar can be shown by setting `legend.text=TRUE`.[9]

4.1.2.2 The Dot Chart

Dot charts are a good alternative to bar plots. Apart from some versions used since the nineteenth century, they were introduced by Cleveland in [6] to show the frequency of categorical values in a variable. The function `dotchart()` evaluated on a vector creates a graph where the values are depicted in the X-axis and the vertical represents the number of times each value is repeated. Should the vector have labels in their coordinates, these are shown in the Y-axis and, if the vector or matrix do not have labels, the argument `labels` can pass names (as a vector of string characters) for each observation in the dot chart.

Dataset `USPersonalExpenditure`, used in Sect. 4.1.1, contains a matrix with the population spending in 5 items for different years. This matrix has the expenditure sectors as row names and the years as column names. The first column inherits these attributes as a vector, the row names being then the labels of the entries. A `dotchart()` of this vector shows the expenditure in the different good and services for year 1940 (see Fig. 4.12, top).

```
dotchart(USPersonalExpenditure[, 1],
         main="US Expenditure Year 1940")
```

However, when evaluated on a matrix, `dotchart()` creates as many dot charts as columns, which allows us to rapidly compare our variables (see Fig. 4.12, bottom).

```
dotchart(USPersonalExpenditure,
         main="US expenditure. Evolution 1940-1960")
```

In Fig. 4.12 top we see the amount of personal expenditure in 1940 with different quantities for each good and service; while in the bottom picture we plot the same values for the whole dataset; therefore, the evolution through years is observed.

4.1.2.3 The Box-and-Whisker Plot

The `boxplot()` function creates a box-and-whisker plot for a data sample. A box plot is an efficient way of depicting the distribution of a sample through its quantiles.[10] The central box shows the interquartile range (IQR), all values between the first and third quartile, and highlights the median with an horizontal line. Outside, two whiskers extend above the third quartile up to 1.5 times the IQR and below the first quartile down to 1.5 times the IQR. The value is 1.5 by default but can be changed with the argument `range`.

---

[9]Legend arguments are passed with `args.legend` and will be explored in detail in Sect. 4.1.3.

[10]This will be explained in detail in Sect. 5.1.1.

US Expenditure Year 1940

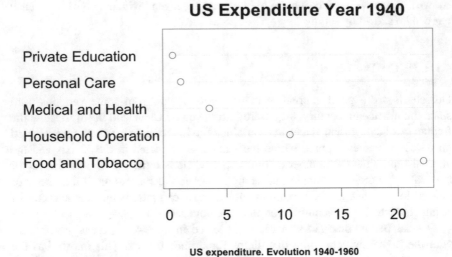

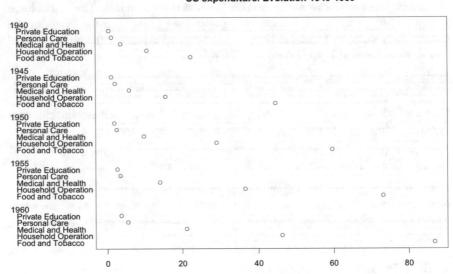

Fig. 4.12 Top: Dot chart of the US personal expenditure in 1940. Bottom: Comparison of the same variables for 5 years of observations

This plot is particularly useful to represent the dispersion of the sample around its median and the skewness with respect to its arithmetic mean. Big boxes come with more disperse data while skewed boxes and whiskers represent data more concentrated in one of the two halves. It can also be understood as a graphical version of the **R** function summary(), providing the main position measures of a distribution. As an example, call boxplot() over one variable of the data frame iris.

```
boxplot(iris$Petal.Width, main="Box plot Petal Width")
```

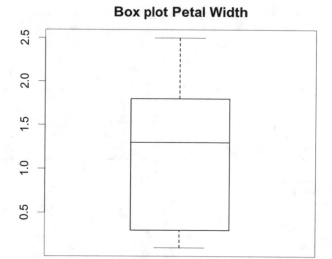

Box plot Petal Width

Fig. 4.13 Box plot of the petal width from dataset `iris`

In Fig. 4.13 we observe at a glance that the median of the petal width is a bit lower than 1.5, but the dispersion is quite big, the 50% of the central values being between 0.3 and 1.7. We see how skewed the distribution of this variable is, the range from the first quartile to the median is much bigger than the range from the median to the third quartile: variability of petal width in the study seems to be higher for lower values.

An interesting additional option of this function is the creation of several box plots in the same image for different categories, introducing another variable in the plot, in a similar way to the usage of variables in arguments in plot(). For example, as we explored in Sect. 4.1.1, the `iris` dataset has observations for three different species. We may want to compare position measures for the three species by simply picturing their respective box plots. To do so, boxplot() should be called connecting iris$Sepal.Length with iris$Species by means of the symbol ~. This way, the boxplot() function separates Sepal.Length into as many subsets as different values of the second variable (the number of distinct species).

```
boxplot(iris$Sepal.Length ~ iris$Species, notch=TRUE,
        main="Box plots Sepal Length for each species")
```

Although it is more suitable for data analysis (Chap. 5), let us make a comment about the argument notch (logical, and FALSE by default). It allows to compare visually whether the medians of any two of the variables are different or not,

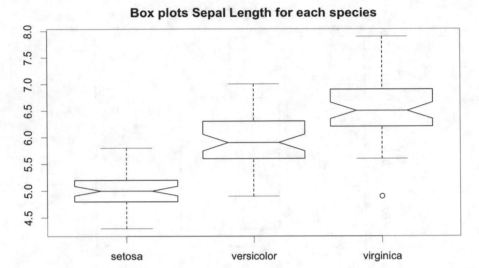

Fig. 4.14 Box plots of the sepal length for the three `iris` species

depending on the notches overlapping.[11] Hence, this example (see Fig. 4.14) shows how, despite some overlapping, the distribution of the sepal length of the virginica flowers is bigger than those of the versicolor, which in turn is bigger than setosa, through box plots, therefore, obtaining a clearer picture than using just the mean or the median. In addition, we observe an outlier depicted as a lower point (of value 4.9 as we will see below), beyond the minimum, for virginica flowers.

Some additional arguments are `width` (numerical, relative width of the box), `boxwex` (also numerical, box scale factor), which are chosen by default to fit on the plot window or `horizontal` (a logical argument set to `horizontal=FALSE` that determines the direction of the boxes). The `border` argument changes the outline of the box plot in a different color.

Advanced plots usually enclose relevant information that can be used but which was hidden at first. Note that, as it is the case with most plotting functions, the output plot can be stored as a variable that can be afterwards consulted for further analysis. In our example we have

```
box.sepal <- boxplot(iris$Sepal.Length ~ iris$Species, notch=TRUE)
box.sepal
```

```
$stats
      [,1]  [,2]  [,3]
[1,]  4.3   4.9   5.6
```

---

[11]The notches are depicted to a distance of ± 1.58 the interquartile range (a dispersion measure of the data explained in Sect. 5.1.2) divided by the square root of the sample size. This calculation, according to [3], gives a 95% confidence interval for the difference between the two medians being statistically significant.

```
[2,]   4.8   5.6   6.2
[3,]   5.0   5.9   6.5
[4,]   5.2   6.3   6.9
[5,]   5.8   7.0   7.9
$n
[1] 50 50 50
$conf
            [,1]        [,2]        [,3]
[1,]  4.910622  5.743588  6.343588
[2,]  5.089378  6.056412  6.656412
$out
[1]  4.9
$group
[1]  3
$names
[1]  "setosa"  "versicolor"  "virginica"
```

The box.sepal object contains all necessary information to build the box plots in a list (recall the notion of list from Sect. 2.2.6). First element, stats, is a matrix with as many columns as produced box plots, where rows correspond to the position measures plotted: end of the whiskers, quartiles, and median comprehending the box. Value n is the sample size producing each box plot, conf contains the limit values of the (confidence interval determined by the) notch, out are the outliers (if any) and group and names are the number and the names of the plotted samples.

4.1.2.4 The Histogram

Many times variables are continuous, taking as values any number, not necessarily an integer.[12] In these situations values are rarely repeated, nevertheless the analyst needs to find the analogous concept to the frequency of a value, the number of times each value appears in the study. For this purpose, grouping values into intervals is a good option. Instead of counting the frequency of a particular value, do it for the amount of times any value in a given interval shows up. Typical examples of this are measuring heights or weights, where (175, 180] represents all values bigger than 175 cm and less or equal than 180 cm, instead of distinguishing between particular values like 176.6 and 177.2 which represent, in a broad sense, the same height for a human person.

The way to graphically represent the frequency of an interval distribution is by means of a *histogram*, which we implement in **R** with the function hist(). Apart from common plot arguments, there are two particular ones in the histogram function. By default, absolute frequencies are shown in a histogram plot and by setting freq=FALSE relative frequencies are plotted instead. Variable breaks is passed as a vector of breakpoints between histogram bars, indicating the interval borders; default is NULL in which case **R** chooses the proper values. An alternative

---

[12] A continuous variable X is a function taking values on the real numbers. See Sect. 5.2.1.

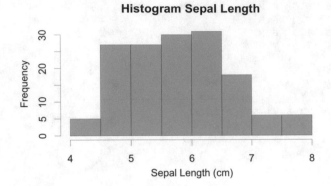

Fig. 4.15 Histogram of the sepal length

way of using `breaks` is to provide a single number that will be understood by **R** as the number of breaks to have, and the corresponding intervals will be automatically computed.[13] Figure 4.15 shows an example of this function with code

```
hist(iris$Sepal.Length, breaks=seq(4, 8, 0.5), col="orange",
     main="Histogram Sepal Length", xlab="Sepal Length (cm)",
     ylab="Frequency")
```

where the values of `Sepal.Length` are divided into 8 intervals of length one half between 4 and 8.[14] With the histogram we can see (Fig. 4.15) at a glance that most of the flowers have sepals of length between 4.5 and 6.5 without distinguishing any further particular values.

As before, `hist()` also provides more information than just the graph. If we store the output of this function in a new object, we can see its content, as in code

```
hist.sepal <- hist(iris$Sepal.Length, breaks=seq(4, 8, 0.5),
                   col="orange", main="Histogram Sepal Length",
                   xlab="Sepal Length (cm)", ylab="Frequency")
hist.sepal
```

```
$breaks
[1] 4.0 4.5 5.0 5.5 6.0 6.5 7.0 7.5 8.0
$counts
[1]   5 27 27 30 31 18   6   6
$density
[1] 0.06667 0.3600 0.3600 0.4000 0.41333 0.2400 0.0800 0.0800
$mids
[1] 4.25 4.75 5.25 5.75 6.25 6.75 7.25 7.75
$xname
```

---

[13]It is important to note that the number of breaks is only interpreted by **R** as a suggestion, so you might ask for `breaks=5` and get a plot with 7 breaks, for example.

[14]Recall that `seq(start, end, by)` creates a sequence vector with the starting and end points and the gap between entries.

```
[1]  "iris$Sepal.Length"
$equidist
[1]  TRUE
attr(,"class")
[1]  "histogram"
```

Inside this object named hist.sepal of class "histogram," we find a list of several items. First, hist.sepal$breaks contains the limits of the intervals and hist.sepal$counts is the absolute frequencies for each interval. In hist.sepal$density we find the relative frequencies (the counts divided by the total number of values) and hist.sepal$mids are the interval midpoints.

4.1.2.5   The Pie Chart

Pie charts are well known graphs used specially with categorical variables where we are interested in measuring relative abundances. If we want to represent vote percentages for different parties in an election, for example, the amount of voters for each party is depicted as a pie slice of the corresponding size. The relevant thing is the relative size of the slices, allowing to compare which values appear at a greater frequency.

This can be done with function pie(). When called over a vector of frequencies, it shows a pie chart where each slice is proportional to the number of times that each category or value appears. Optionally, argument labels can add a vector of labels for each slice; when not provided, **R** uses the numerical values of x as labels. Apart from that, radius indicates the circle radius, taking the value 0.8 by default.

For example, let us get advantage of data from Fig. 4.15 stored in the variable hist.sepal. The following code plots (see Fig. 4.16 left) a pie chart with a slice for each histogram bar, labeling them with the midpoint of the interval as in

```
pie(hist.sepal$counts, labels = hist.sepal$mids, col=1:8,
    main="Sepal length distribution")
```

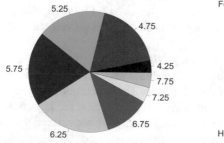
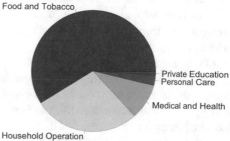

Fig. 4.16 Pie chart of the US Personal Expenditure in 1940

Apart from using `pie()` over a vector of frequencies, it can be used over a numerical vector itself. In that case, each slice of the pie chart is proportional to the value of the vector. For example, this code plots a pie chart of the personal expenditure for the first column of `USPersonalExpenditure`, yielding rapidly that food and tobacco were the most consumed items, with more than half of the personal budget (Fig. 4.16 right)

```
pie(USPersonalExpenditure[,1], col=rainbow(5),
    main="US Personal Expenditure in 1940")
```

where we have used `rainbow(n)` for slice colors, showing a different color (following a rainbow sequence) for each value.

4.1.3   Customizing Plots

Previously, we have seen a variety of possibilities to represent data graphically. Once we have created our plots, **R** allows to elaborate more on the final design. We can show a legend telling the series of data plotted or add text to a plot in different ways. Besides, we can distribute the plot or plots in the window frame modifying the screen layout. Finally, we will learn how to save our plots for later use in presentations and communications of our study.

4.1.3.1   Adding Text to a Plot

There are many reasons why we might be interested in adding text to a plot. For example, data groups are plotted in different colors and we want to advice which color corresponds to what. Or some region in the picture deserves a particular text label indicating an interesting feature to look at, an insight coming from the analyst, not from the data.

In previous sections we have seen how to include axis and tick labels, which is an essential information when plotting coordinate axis, to show the variables depicted and the scale of them. Also, a plot title has to contain a brief description of the plot. But sometimes we need more than that and extra characters are necessary. We will see how we can do it with functions `legend()` and `text()` from package `graphics`. Function `mtext()` will add main texts to array plots created with `par()`.

One of the most important things when analyzing data is to clarify the meaning of the points and lines that we represent graphically. This leads to the notion of *legend*, which can be added to a plot by the `legend()` function. The first arguments are `x` and `y`, the coordinates where the text is written; although we can pass the keywords `"bottomright,"` `"topleft,"` `"right,"` etc., as `x`, and leave the default `y=NULL` to locate the legend in a distinguished position in the picture. The argument `legend` is exactly the text to be written; should several series of data to be plotted,

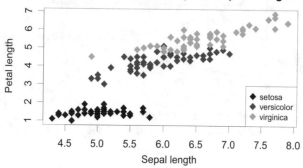

Fig. 4.17 Figure 4.5 with legend

a vector of strings will allocate the text legend to each series. When the legend refers to different data series, arguments pch, lty, lwd, col, and cex can specify a vector of point types, line types, widths, and colors for each series (if a single value is introduced, it applies to all series). Box size can be changed with cex and points size inside the box legend is changed with pt.cex. Legend box can be removed by setting bty to "n" (default is "o").

We now add a legend to the right plot in Fig. 4.5 to obtain Fig. 4.17. We locate the text in the bottom right corner of the picture and add the text of the three data categories by means of the levels of the variable Species (the one we used in Fig. 4.5 to distinguish colors). Colors in the plot are assigned black, red, and green by default (numbered colors 1, 2, and 3 in **R**) and so is the legend. We also set the legend point type and point size to the same as in the plot (pch=18 and pt.cex=1.5) for consistency.

```
plot(matrix.iris[, c(1, 3)], pch=18, cex=1.5, col=iris$Species,
     main="Comparison between sepal and petal length",
     xlab="Sepal length", ylab="Petal length")
legend("bottomright", legend=levels(iris$Species), col=1 : 3,
     pch=18, cex=0.8, pt.cex=1.5)
```

It is worth noting again that, when a plot is displayed, that plot is an active object in **R** subject to adding new elements. This happens with legend; once a plot is created, calling the legend adds this element to the picture. The same can be done with text(), which introduces an additional text in the graphic. The command text(x,y,labels) shows a string character in a point of given coordinates (x, y) where labels contains the text to be written. We can specify the position where the text refers to, but inserting the text on the bottom, left, top, or right to that position with pos=1,2,3,4. The argument font is a number specifying the type font for the text, where 1 corresponds to plain text (default), 2 to bold face, 3 to italics, and 4 to bold italics.

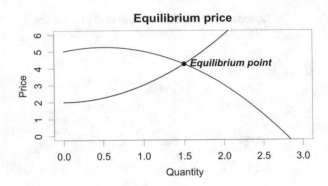

Fig. 4.18 Equilibrium point in a supply–demand diagram

An example can be produced by adding a text to the plot in Fig. 4.9, specifying the equilibrium point when the supply and demand curves intersect.

```
curve(supply, from=0, to=3, ylim=c(0, 6), lwd=2, col="blue",
      main="Equilibrium price", xlab="Quantity", ylab="Price")
curve(demand, add=TRUE, lwd=2, col="red")
points(x=1.5, y=4.25, pch=19)
text(x=1.5, y=4.25, labels="Equilibrium point", pos=4, font=4)
```

In Fig. 4.18 we have the price of supply and demand as functions of the quantity. The equilibrium price is the point where supply matches demand, and this happens when a price equals to 4.25 (and a quantity of 1.5). After plotting the supply curve, with the plot being active, we show the demand curve and insert a point with the given coordinates with `points()`. Then, add the corresponding text at the right of the given position, in bold italics.

4.1.3.2 Graphical Parameters

Once we have a clear idea on how we want our plot to look like, to visualize the whole information at a glance, we may need to modify the size of the margins or include several separated graphs into the same plot. These *environmental features* can be customized with functions `par()` and `layout()`.

Frequently, we are interested in plotting several images in the same graph, showing different characteristics of the data to study. The function `par()` allows to distribute *multiple graphs* and margins into the same plot. When typing `par(mfrow=c(a,b))`, it creates an $a \times b$ array of plots, filled by rows with $a \times b$ plots called afterwards (argument `mfcol` does it in a similar way, filling the array by columns).

This function also features non required arguments such as `mar` and `oma`, which accept numerical vectors of four. These values set the bottom, left, top, and right

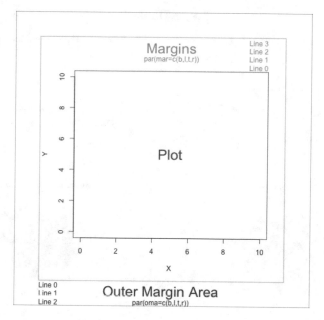

Fig. 4.19 Outer and inner margins in a **R** plot

inner and outer margins, respectively, in number of lines of text, see Fig. 4.19[15] for an example.

A very typical situation is when we pursue to build a plot containing some of the graphs that we have previously created. For example, we can gather in the same picture all the visual information about sepal lengths from the iris dataset that we collected in Figs. 4.17, 4.14, 4.15, and 4.16. We create a 2 × 2 array of images with par (mfrow=c(2, 2)). When titles are not used, graphs may have blank regions around them, so we can reduce the inner margins, for example, with par (mar=c(3,3,2,2)). It is generally wise to leave wider gaps for the bottom and left margins in case we want to use axes titles. The plot still lacks a common main title for the four pictures. First, we create room for it by adding the argument oma=c(0, 0, 2, 0) which leaves 2 lines of text at the top of the picture, where we can write the title.

The main title *The Sepal Distribution* is added with the function mtext() which introduces a main title on a multiple plot. The arguments are text, which is the title to be written and side, which controls the position of the plot where we want the text to appear (1 for bottom, 2 for left, 3 for top—default—, and 4 for right).

Finally, set the flag outer to TRUE if outer margins are available; otherwise the main title will be posted in the last array plot.

---

[15]Image provided by Holtz [9] via https://www.r-graph-gallery.com/74-margin-and-oma-cheatsheet/.

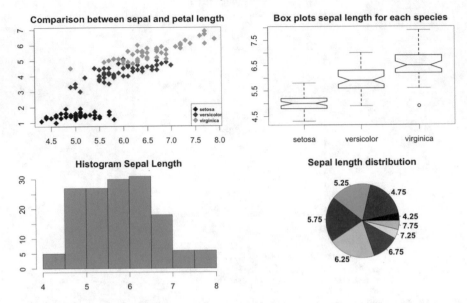

Fig. 4.20 Array of sepal length plots created with par

The following code generates Fig. 4.20, where we have set `side=3` to have the title on top, `outer=TRUE`, and added `font=2` so that titles appear on bold.

```
par(mfrow=c(2, 2), mar=c(3, 3, 2, 2), oma=c(0, 0, 2, 0), font=2)
plot(matrix.iris[, c(1, 3)], pch=18, cex=1.5, col=iris$Species,
     main="Comparison between sepal and petal length",
     xlab="Sepal length", ylab="Petal length")
legend("bottomright", legend=levels(iris$Species), col=1:3,
       pch=18, cex=0.6, pt.cex=1.5)
boxplot(iris$Sepal.Length ~ iris$Species, notch=TRUE,
        main="Box plots sepal length for each species")
hist(iris$Sepal.Length, breaks=seq(4, 8, 0.5), col="orange",
     main="Histogram Sepal Length",
     xlab="Sepal length (cm)", ylab="Frequency")
pie(hist.sepal$counts, labels = hist.sepal$mids,
    col=1 : 8, main="Sepal length distribution")
mtext("The sepal distribution", side=3, line=1, outer=TRUE)
```

Further ways to express information include rescaling the graphs, for example, with a bigger scatterplot and a smaller pie chart in Fig. 4.20. We can combine plots of different shapes and sizes into a single window using argument `fig` in `par()` or by using the function `layout()`. We present only the latter as it is more versatile, but the former can be explored in Exercise 4.8.

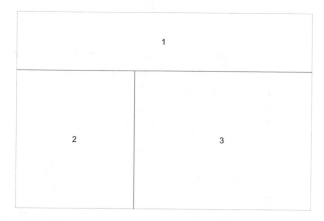

Fig. 4.21 Layout window divided into plot boxes

We now gather the plots we created with the USPersonalExpenditures dataset. Imagine that we want to display three plots in the same window, where the plot in Fig. 4.11 occupies the top part whereas the bottom part is occupied by Figs. 4.16 left and 4.12 bottom, and the second uses most of the lower area. This way, we have three plots to be included in a layout frame as shown in Fig. 4.21 below. In order to encode this visual information we create a matrix for the layout that is filled with the plot numbering that identifies what part of the layout hosts each plot. Since we have two rows and two columns we use a 2 × 2 matrix. Both entries of the first row of the matrix should be filled with number 1, as the first row features only that plot. The second row is filled with numbers 2 and 3.

The layout() function is now called over this matrix, specifying the relative widths and heights of each sub box in the window. If no widths and heights are specified, equal-sized boxes are displayed. The following code

```
lay.matrix <- matrix(c(1, 1, 2, 3), nrow=2, ncol=2, byrow=TRUE)
layout(lay.matrix, widths=c(0.4, 0.6), heights=c(0.3, 0.7))
layout.show(n=3)
```

creates an upper box of $0.3 = 30\%$ height size and two lower boxes of the remaining $0.7 = 70\%$ height and relative 0.4 and 0.6 widths. The result is shown in Fig. 4.21, called with the command layout.show(n) that shows the empty layout, where n is the number of plots of the total window to be shown.

The following code first produces the layout as above, then calls par() to set the desired margins so as not to overlap graphical elements and finally draws the three plots adding a main text on top with mtext(side=3), in italics (font=3). The resulting graph can be seen in Fig. 4.22.

```
lay.matrix <- matrix(c(1, 1, 2, 3), nrow=2, ncol=2, byrow=TRUE)
layout(lay.matrix, widths=c(0.4, 0.6), heights=c(0.3, 0.7))
par(mar=c(2, 2, 2, 1))
barplot(USPersonalExpenditure[,5], ylab="Billions of dollars",
        main="Expenditure  in 1960", col="darkgreen")
```

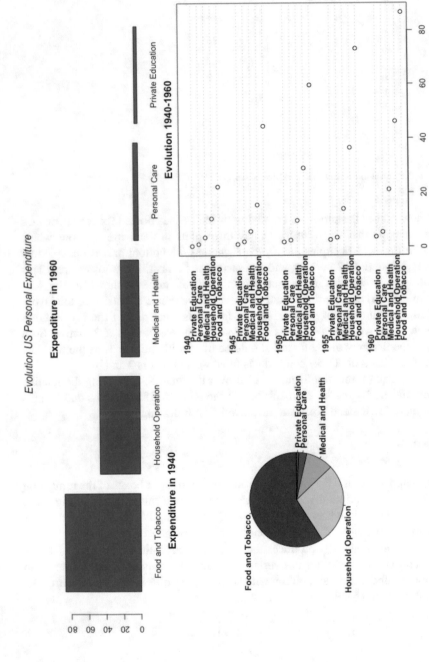

Fig. 4.22 Plot array of USPersonalExpenditures with layout function

```
pie(USPersonalExpenditure[,1], col=rainbow(5),
    main="Expenditure in 1940", radius=0.5)
dotchart(USPersonalExpenditure, main="Evolution 1940-1960")
mtext("Evolution US Personal Expenditure", side=3, outer=TRUE,
    font=3)
```

An elaborated distribution of our plots creates a much more powerful graph, where all the relevant information is synthesized into several ways of visualization. Arranging margins and plots in the same window comes at the price of some time spending; despite this, the quality of the output deserves it.

So far we have discussed how to display several plots in the same window. If, instead we wish to visualize plots in different windows, we can show new ones with command dev.new(). This will create a new window where a new plot will be located without removing the old one. The opposite is function dev.off(), which closes the current active graphic device. Care should be taken when trying functions like par() or layout() because **R** might not respond as we expect to our plotting commands, due to a previously opened device with defined parameters. We can simply clear them with dev.off().

4.1.3.3 Saving a Plot

Saving our work is essential with any software. When creating images with **R** we can save them in different format types, including .bmp, .jpeg, .png, .svg, .pdf, and .tiff formats. Each one of them serves for different purposes. For general use, .jpeg and .png are the most suitable ones due to its trade-off between quality and size. The vector format .pdf is recommended when integrating our plots into another document because it preserves all distances without losing definition. Format .svg is a vector image format that allows interactivity and escalation without quality loss.

We will save our plot by calling special instructions before and after creating the plot. The starting point is to call the format we want: bmp(), jpeg(), png(), svg(), or pdf() with the name of the file that the image will be stored as, in quotation marks. To this, we can include arguments width and height with the dimensions of the image (both equal to 7 inches by default), or specify the pointsize. This opens an object with that format in the directory to be saved.

Once all the elements that we want to show in the plot have been added (including the possibility of an array of plots, adding texts, legends, etc.), we will create our image calling dev.off(), which closes the opened object. Notice that when an image format function is called, no graphs are displayed in RStudio.

```
png("Iris-histogram.png", width=400, height=300)
hist(iris$Sepal.Length, breaks=seq(4, 8, 0.5), col="orange",
     main="Histogram Sepal Length", xlab="Sepal Length (cm)",
     ylab="Frequency")
dev.off()
```

The folder where the plot is saved is the current working folder of our **R** session. To know which one is it, call getwd(). To change the working directory into a new one, call setwd() with the new directory. This can also be done by clicking in Files and More in the bottom right window of RStudio. In RStudio, plots can also be easily saved as well through the button Export in the *Plots* menu and then choosing either Save as Image (where format and dimensions can be chosen) or Save as PDF.

4.1.4   Exercises

Exercise 4.1 An interesting dataset in **R** is trees. It contains information about 31 black cherry trees, which have been measured in girth, height, and volume. Answer the following questions using the dataset and the visualization techniques seen in this section.

(a) When girth is increased, which other variable is more likely to grow?
(b) What variable shows systematically higher values?
(c) How many mode values do you see in the distribution of the variable volume?
(d) Which one of the three variables seems to have a more skewed distribution?

Exercise 4.2 Another interesting dataset is rock, which contains measurements on 48 petroleum rock samples. The variables in the frame are area of the pores (area), total perimeter of the pores (peri), shape (shape), and permeability (perm). Using your current knowledge answer the following questions:

(a) Analyze the scatterplots of the given variables. Do you find any relations?
(b) Do scatterplots for the following pairs of variables: area vs peri, peri vs shape, and shape vs perm. All plots seem to show two main groups of individuals. Check if all groups are made of the same individuals. What are your conclusions?
(c) Use a histogram to obtain a clear evidence of the presence of two distinct groups in the population.
(d) Repeat the histogram using this time relative frequencies and connect the top of the bars with straight lines.

Exercise 4.3 Plot the function $f(x) = x^2$ between $x = -1$ and $x = 3$ to reproduce the left image in Fig. 4.23. To do that,

1. Define $f(x)$ as a **R** function.
2. Evaluate $f(x)$ at points $-1, 0, 1, 2$, and 3.
3. Plot them as a scatterplot and then connect the points with lines.
4. The picture looks like some consecutive straight lines. How could you increase the detail of this plot to show that it is actually a curve?

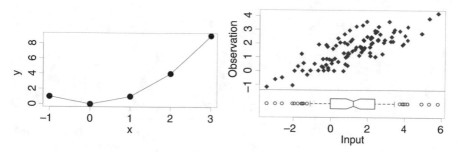

Fig. 4.23 Images corresponding to Exercises 4.3 and 4.4

Exercise 4.4 Reproduce the image in Fig. 4.23 right by plotting a certain random sample. Random samples and distributions will be studied in detail in Chap. 5, but for now type the following code:

```
set.seed(1)
x <- rnorm(100, 1, 2)
y <- 0.5 * x + rep(1, 100) + rnorm(100, 0, 0.7)
```

As you see, this generates 100 observations of two variables, x and y that will be used to create several plots.

1. Create a scatterplot of y vs x with blue diamonds of size 1.5 as the marker. Name axis Y "Observations" and remove axis X.
2. In a separate plot call, create a horizontal box plot for the variable x, with whiskers ranging 50% of the interquartile box size. Add a notch and red borders of width 2. Remove the Y-axis and name the X-axis as "Input".
3. Combine both plots in a single graph using the bottom 40% of the plotting window for the box plot. Hint: call function par() twice to remove the necessary margins.

Exercise 4.5 We now inspect the airquality dataset. Start plotting ozone levels (Ozone) vs. temperature (Temp) as in the left image of Fig. 4.24 with the following instructions:

1. The size of the bubble points is a third of the wind level for that observation.
2. The color of the marker is different for every month.
3. Axes range from −5 to 170 horizontally and from 50 to 100 vertically.

Exercise 4.6 Now plot the evolution of the temperature (Temp) as in Fig. 4.24 right, with the following specifications:

1. Temperature is represented as a thick line over time.
2. Time goes from first observation (1) to last (153).
3. Triangle points are drawn just for those days when ozone values are larger than the temperature measurement for that day.
4. No gap is left for top or right margins.

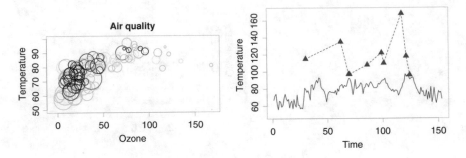

Fig. 4.24 Images corresponding to Exercises 4.5 and 4.6

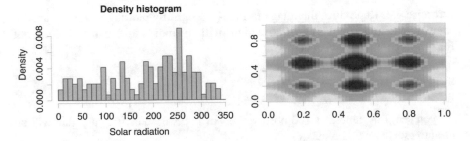

Fig. 4.25 Images for Exercises 4.7 and 4.8

Exercise 4.7 Reproduce the histogram depicted in Fig. 4.25 left, using the column of the solar radiation variable (`Solar.R`) from the `airquality` dataset. For that,

1. Bars are filled with color gray and borders are red.
2. Histogram represents relative frequencies.

Exercise 4.8 The egg crate function (see below) is a complex three dimensional function difficult to visualize. The *isopleths* or contour lines can be used to visualize the altitude of a function at a given location (x, y).

$$f(x, y) = x^2 + y^2 + 25(sin^2(x) + sin^2(y)) \qquad (4.1)$$

1. Evaluate the function in the (X, Y) square $[-5, 5] \times [-5, 5]$, and save its values in a 50×50 matrix. Then, investigate on the Internet how to represent this kind of data and visualize the stored matrix using contour lines. Hint: you may use function `seq` with argument `length` to generate the X and Y values.
2. An alternative is to use function `image()`. Use it to represent the same matrix and search on the help files how to modify its colors (function `image()` rescales the matrix and displays a picture with axes between 0 and 1). Try to reproduce Fig. 4.25 right.

4.2 The `ggplot2` Library

When data visualization requires of more advanced or elaborated graphics, other packages can be used. Sometimes the use of more proficient images comes from the need of sharing a beautiful display in order to effectively broadcast the results of a study, other times the need is intrinsic, for example, if maps are needed, or when overlaying different types of plots in a way that exceeds `plot()` capabilities. One of the most commonly used data visualization packages is `ggplot2`,[16] written by Hadley Wickham, see [15], and based on the visualization philosophy of Leland Wilkinson, see [17].

This package is part of `tidyverse`, a collection of **R** packages that *share an underlying design philosophy, grammar, and data structures* for every step in data manipulation: importing, tidying, transforming, modeling, and visualizing, see [14, 16]. This assortment of libraries was a breakthrough in the **R** ecosystem, featuring a wide range of new tools and became extremely popular. Of late, though, it has fallen behind `data.table` for data preprocessing as the latter has included most of the features of `tidyverse` with an equally simple syntax but with incredibly faster and more efficient results. That is the reason why in this book we only cover `ggplot2` among the several packages included in `tidyverse`.

In Sect. 4.1, we saw multiple **R** functions for different kinds of plots: scatterplots, box plots, histograms, images, etc. Although many arguments are similar, each function has its own syntax, limiting its versatility and frequently making it difficult to produce elaborated graphics. With `ggplot2` we can integrate all our plots into a single common grammar, where we can define all possible arguments and expand the plotting possibilities far beyond the limits of the `graphics` package. Therefore, `ggplot2` can be understood as a structured programming language on its own, with objects classified into types and subjected to strict grammar rules. This may seem excessive when we just need a quick glance to our data, but for more elaborated plots to be made public, the effort is worthy and `ggplot2` is advisable.

To load the package, we proceed as usual.

```
library(ggplot2)
```

It is beyond the scope of this book to detail the entire package, as it accounts for hundreds of functions.[17] Nevertheless, we will not only focus on the main functions and elements allowing to produce beautiful outputs, but we will accompany them with interpretations in order to show how insight can be gained through visualization.

---

[16]The ggplot2 motto is *Create Elegant Data Visualizations Using the Grammar of Graphics*.

[17]For more examples and full description of all functions, visit https://ggplot2.tidyverse.org/index.html.

4.2.1   Grammatical Layers

The syntax of all ggplot2 commands is quite different to the code we have
studied so far but it is still intuitive, nonetheless. The whole philosophy behind it
lays in adding layers of content that stack up to the final plot. The main function
for graphing with this library is ggplot(). This function has two required
arguments,[18] data which is the dataset to be used for the plot (which has to be
formatted as a data frame) and mapping=aes() which controls the variables to
use. This second argument is always passed by means of aes(x, y), a function
that selects the variables inside the data frame to be plotted. The name aes comes
from *aesthetics*, which are the properties of the plot that we create.

The ggplot() function, though, will not graph anything by itself, it just loads
an environment focused on a particular dataset where more information can be
added afterwards. For example, the code

```
ggplot(iris, aes(x=Sepal.Length, y=Sepal.Width))
```

displays a gray rectangle with a white grid, as if it was a canvas waiting to be
filled with graphics. Note that we are using iris, the built-in dataset introduced
in Sect. 4.1.1.

We are now ready to add layers of content to the empty canvas. Instead of passing
them as arguments of the ggplot() function, they are defined as standalone
functions and are added one by one with the + symbol.

Apart from the basic ggplot(data, mapping) layer that sets the environment, there is another mandatory layer, the geometry layer. The most basic geometry
layer is geom_point(), a function that adds points corresponding to the variables
defined by aes(). Thus, the code

```
ggplot(iris, aes(x=Sepal.Length, y=Sepal.Width)) +
  geom_point()
```

produces the plot in Fig. 4.26. It is a long and complicated code for something
we could easily perform with plot() but now stacking other figures, or adding
enhancements will be easier and more natural. The rich topic of all different
geometry layers that can be added is discussed in the following section.

4.2.2   Geometry

The geometry layer is the last necessary layer to produce a finished plot in
ggplot2. Many different geometry functions are available for a variety of
plots that clearly widens the possibilities of plot(). Table 4.1 shows the most

---

[18]It allows more than the two required arguments, but their purpose can be achieved in a more
natural way with other layers.

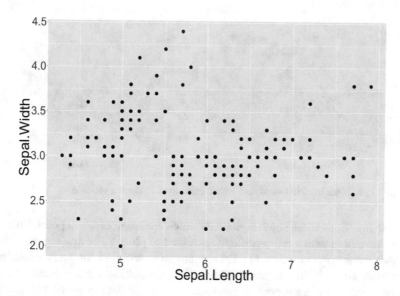

Fig. 4.26 Scatterplot of the sepal length and width with `ggplot2`

Table 4.1 Common geometry `ggplot2` functions with definition and specific elements

Function	Use	Specific elements
`geom_point()`	Scatterplot	
`geom_abline()`	Straight line	`slope, intercept`
`geom_hline()`	Horizontal line	`yintercept`
`geom_vline()`	Vertical line	`xintercept`
`geom_line()`	General line	`linetype`
`geom_smooth()`	Estimated tendency	`method, formula, se, n` `span, fullrange, level`
`geom_bar()`	Bar counts	`width`
`geom_col()`	Bar diagram	`width`
`geom_histogram()`	Bar histogram	`binwidth, bins, breaks`
`geom_freqploy()`	Polygon histogram	`binwidth, bins, breaks`
`geom_density()`	1D density estimation	`bw, adjust, trim`
`geom_errorbar()`	Vertical error bars	
`geom_boxplot()`	Boxplot	`notch, notchwidth,` `varwidth, outlier`
`geom_contour()`	Contour plot	`lineend, linemitre`
`geom_density_2d()`	2D density estimation	`lineend, linemitre` `contour, n, h`

commonly used geometric functions, together with a brief definition and some of the specific elements (when existing). These functions cover most of the plot types seen in Sect. 4.1 and add some more.

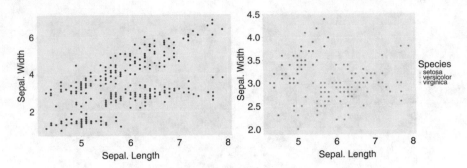

Fig. 4.27 Left: color used in two layers. Right: color used through a variable

These functions share some arguments which are common to all possibilities and have some others which are specific to each kind of plot.[19] The first argument is the already seen mapping=aes(). This argument is required for every geometrical layer; after all, it is the command that tells the function what variables to plot. Nevertheless, if no argument is provided, as in the last code of the previous section, the mapping defined in the ggplot() is inherited. This has an important consequence, aes() might not be defined in the original ggplot() call, but in each geometry layer so that each layer might use different variables. For example with

```
ggplot(iris) +
  geom_point(aes(x=Sepal.Length, y=Sepal.Width), color="red") +
  geom_point(aes(x=Sepal.Length, y=Petal.Length), color="blue")
```

produces the left-hand side of Fig. 4.27. Note that here we used the argument color which together with size, shape, fill, stroke, and alpha[20] are secondary arguments to control aesthetic elements. These arguments can be used both outside or inside of the aes() command but we have to be extremely careful here as the results are different doing one way or another.

When these arguments are used outside aes() then **R** is expecting a value for each argument and the desired output is as above. However, if the arguments are specified inside, then **R** is expecting a variable (from the data frame) that will provide the way of using the argument. For example, the code

```
geom_point(aes(x=Sepal.Length, y=Petal.Length, color="blue"))
```

will not produce a blue graph. The correct way of using it is passing a variable of the data frame that will drive the usage of the color. Instead, the code

```
ggplot(iris) +
  geom_point(aes(x=Sepal.Length, y=Sepal.Width, color=Species))
```

---

[19]The main specific arguments are listed in the table and exploring them is left to the reader since, by now, it should be straight forward.

[20]The name *alpha* is a standard way to refer to transparency, not only in programming but also in picture or video edition.

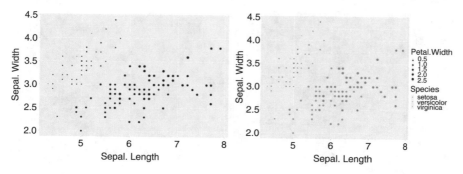

Fig. 4.28 Left: `Petal.Length` as point transparency. Right: `Species` as colors and `Petal.Width` as point size

will produce the right picture in Fig. 4.27 which is the reconstruction of the right picture in Fig. 4.5.

Therefore, when using visual customization arguments outside `aes()`, a specific value should be provided, whereas when using it inside we should provide another variable that determines the usage. The following code explores similar uses of this tool, where we combine the scatterplot of sepal width against sepal lengths, together with petal length transparency scale (Fig. 4.28 left) and petal width scale for point sizes with three colors for the different species (Fig. 4.28 right). This allows to group together many different but related pieces of information in the same plot.

```
ggplot(iris) +
  geom_point(aes(x=Sepal.Length, y=Sepal.Width,
                 alpha=Sepal.Length))

ggplot(iris) +
  geom_point(aes(x=Sepal.Length, y=Sepal.Width, color=Species,
                 size=Sepal.Width))
```

Recall from Sect. 4.1 that we introduced the idea of *representing more than two variables in a two dimensional plot* by means of aesthetic properties such as color, and now size and transparency. In the right part of Fig. 4.28 four variables are represented. Not only they are somehow depicted, but conclusions about the relations between the four of them can be drawn by an inspection of the graph although our eye should be instructed.

Examine first the variables pairwise. If we just look to the sepal length and width, we perceive no relation between them. It is difficult to free your mind from the color, but go back to the left plot in Fig. 4.5 if need be. The distribution of the points in the vertical axis is similar independently of the position in the horizontal axis. Note however that the picture suggests a relationship between sepal length and petal width. As the first increases (go right in the plot), the second does as well (points are bigger in general). Sepal length is also related to species; setosa is located between 4 and 6, versicolor spreads from 5 to 7, and virginica stays (mainly) in the range from 6 to 8. A connection can be established also between the two variables represented

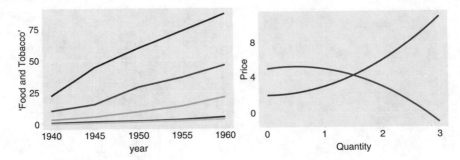

Fig. 4.29 Left: USPersonalExpenditure dataset. Right: supply and demand plot

by properties instead of geometric dimensions as the petal width is smaller (smaller size of the dot) for setosa species (in red). Finally several variables can be taken into account at once. Take a look again at the sepal length and width, this time keeping the color in mind. Even though no relation was hinted between length and width in general, it is clear that in each specie, as the sepal length grows, the sepal width does as well.

Notice the amount of information that is effectively condensed in just one plot. Here the visualization becomes crucial as now, when we learn how to prove mathematically that there is a relation between two variables, we know in advance that we will find such relation only if we subset by species, an important fact that might have been overlooked otherwise.

4.2.2.1 Lines

The strength of ggplot() starts to show off when several layers are added to the main one. Recall that in order to draw a line graph with the basic plot() function we had to set an specific value to the argument type. The concept with ggplot() is slightly different and every single element of a plot has its own command. The function for drawing a curve is geom_line() which shares the main arguments with geom_points() but links the given values.[21]

The basic thing to keep in mind when plotting graphs in this way is that a data.frame must be passed. The following two pieces of code create the two plots in Fig. 4.29, which, respectively, reproduce Figs. 4.7 and 4.9.[22]

```
US.frame <- as.data.frame(t(USPersonalExpenditure), row.names=F)
US.frame <- cbind(year=c(1940, 1945, 1950, 1955, 1960), US.frame)

ggplot(US.frame) +
  geom_line(aes(x=year, y='Food and Tobacco')) +
```

---

[21]The line type, width, and other components that relate to the particular aspects of a line can be modified by using several secondary arguments that the reader can check in the documentation.

[22]Except for the main title and axes labels.

```
    geom_line(aes(x=year, y='Household Operation'), color=2) +
    geom_line(aes(x=year, y='Medical and Health'), color=3) +
    geom_line(aes(x=year, y='Personal Care'), color=4) +
    geom_line(aes(x=year, y='Private Education'), color=5)
```

```
supply <- function(x){x ^ 2 + 2}
demand <- function(x){5 + x - x ^ 2}

values <- seq(0, 3, 0.1)
supply.demand <- data.frame(Quantity=values, Price=supply(values),
                            Price2=demand(values))

ggplot(supply.demand) +
  geom_line(aes(x=Quantity, y=Price), color=4) +
  geom_line(aes(x=Quantity, y=Price2), color=2)
```

Note that, in the first code we transform the original matrix into a data frame (after transposing with function t() to have the expenditures as variables and years as observations), and we need to add the corresponding commas to each y in the geometric lines, to achieve the syntax of the data frame. We proceed likewise with functions in the second example, creating a data frame of supply and demand values. This makes the code look worse, or at least longer, than the original one, but if we had a data frame from the beginning, then the last lines are really easy to follow and very intuitive. Keep in mind that when doing data analysis we are barely always handling data frames and, when not, creating them is not that hard, one or two lines of code that are justified by the versatility and quality of ggplot().

In many plots it is valuable to have the possibility of adding straight lines. Sometimes lines have a deep intrinsic meaning, such as the regression line in the linear model to study the relation between two or more variables (which will be introduced in Sect. 5.3.2). Other times we simply want to stress the separation of two regions or highlight a threshold. Straight lines are easily added in our plots with the functions geom_hline(), geom_vline(), and geom_abline(), depending on whether they are horizontal, vertical, or otherwise, respectively. In order to define the line in each case, the arguments xintercept, yintercept or intercept, and slope are used.

Going back to the iris dataset, we plot two examples with straight lines. From the plots in this section and in Sect. 4.1, it is clear that the size of setosa flowers is considerably smaller than the other two species. The first code separates the setosa group in the scatterplot of petal widths against lengths, by drawing a vertical line at $x = 2.5$ and a horizontal one at $y = 0.75$. In the second, the linear regression line is graphed on the scatterplot.[23] The output is displayed in Fig. 4.30.

```
ggplot(iris) +
  geom_point(aes(x=Petal.Length, y=Petal.Width, color=Species)) +
  geom_vline(xintercept=2.5, lty=2) +
  geom_hline(yintercept=0.75, lty=2)
```

---

[23]The calculations for the slope and intercept will be studied in Sect. 5.3.2.

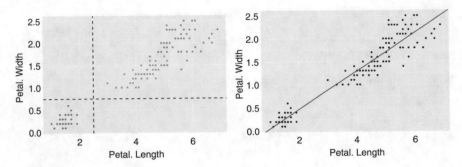

Fig. 4.30 Left: `iris` scatterplot with vertical and horizontal thresholds. Right: same plot with the regression line

```
ggplot(iris, aes(x=Petal.Length, y=Petal.Width)) +
   geom_point() +
   geom_abline(intercept=-0.3630755, slope=0.4157554,
               color=2,lty=1)
```

4.2.2.2   Other Layers for Univariate Series

This subsection shows the `ggplot2` counterpart to Sect. 4.1.2. Since the use of these kind of plots was already explained there, we quickly overview the equivalent commands without looking into all the details, assuming that the reader is already capable of guessing some of the arguments.

Bar plots are now created with `geom_col()` and `geom_bar()`. The first command, equivalent to `barplot()`, takes a numerical vector and plots a series of bars of the corresponding heights, whereas the second takes a categorical variable and counts the number of observations.

In order to fully appreciate the use of commands, consider the `diamonds` dataset from `ggplot2` package as an example, which is a large and interesting sample containing the characteristics of over 50,000 diamonds, measured in 10 variables, some of them continuous and some categorical.

The following code

```
ggplot(diamonds) + geom_bar(aes(x=clarity))

ggplot(diamonds) + geom_bar(aes(x=clarity, fill=color), width=1)

ggplot(diamonds) + geom_col(aes(x=clarity, y=price, fill=color))
```

creates the three plots in Fig. 4.31. To the left, we see the plain bar plot, where we appreciate that types `SI1` and `VS2` are the most abundant ones. The middle plot features a fill coloring based on the color of the diamonds, providing additional

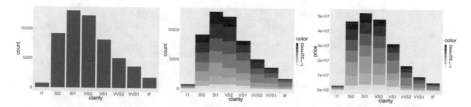

Fig. 4.31 From left to right: bar plot of the diamonds clarity, diamonds clarity per color type, and sum of total price per clarity type

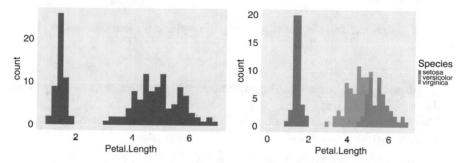

Fig. 4.32 Left: histogram of the petal length. Right: same histogram with different binning per flower type

insight. The third plot does the same for the numerical variable `price` using `geom_col()`.

When dealing with continuous variables, `geom_histogram()` is the correct alternative to bars. These functions work similarly to function `hist()` from Sect. 4.1.2, where `binwidth` and `bins` are the width and number of the bins, and `breaks` is the vector of breaks between intervals.

Default values are chosen by `ggplot2` depending on the range of values. Using again the `iris` dataset, the following code first shows a histogram of the variable `Petal.Length` and, then, additional arguments are included, to color by species and make custom breaks (Fig. 4.32).

```
ggplot(iris) + geom_histogram(aes(x=Petal.Length))

ggplot(iris) + geom_histogram(aes(x=Petal.Length, fill=Species),
                    breaks=seq(0, 7, 0.2))
```

Box plots were already addressed in Sect. 4.1.2, where we defined them as an easy way to visualize the variable distribution. In `ggplot2`, this kind of plot can be performed by using the function `geom_boxplot()`. Similarly to `boxplot()` in package `graphics`, the default value for the whiskers size is again 1.5 times the size of the interquartile range (IQR). A `notch` can be displayed (and customized using the `notchwidth`) to visualize the hypothesis test comparing different medians. In addition to the notch specifications and the usual aesthetics elements,

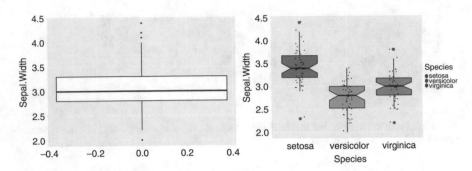

Fig. 4.33 Left: box plot for a single variable. Right: comparison between three variables in box plots

the appearance of the outliers can be customized[24] to enrich graphics in comparison to the original `boxplot()`.

The following two lines of code create a basic and a rather customized box plot that can be seen in Fig. 4.33.

```
ggplot(iris) + geom_boxplot(aes(y=iris$Sepal.Width))

ggplot(iris, aes(x=Species, y=Sepal.Width, fill=Species)) +
  geom_boxplot(notch=TRUE, outlier.color=6, outlier.shape=16,
               outlier.size=3) +
  geom_jitter(size=0.3, width=0.1)
```

The right-hand side image has the extra layer `geom_jitter()` that displays the points over the box plot with small variations in the abscissa. Additionally, note that the `mapping=aes()` is not passed in each layer but in the main `ggplot()` call. This is considered a good practice when all the layers have the same input variables. In such cases, we save time and make the code more clear by passing `aes` at the very beginning.

The graph in the right-hand side shows a consistent result to what was depicted in Figs. 4.27 and 4.30. Both versicolor and virginica species have overlapping distributions, a challenging scenario when trying to differentiate them, whereas setosa is clearly separated even when checking only one of the variables.

4.2.2.3   Uncertainty and Errorbar Layers

In data analysis, looking for *relationships* between variables is a constant problem. For example, the regression line in the right plot of Fig. 4.30 describes a linear relation between the two variables. In general, as the length of the petal increases, so

---

[24]By means of `outlier.color`, `outlier.fill`, `outlier.shape` (which hides the outliers if set to NA), `outlier.size`, `outlier.stroke`, and `outlier.alpha`.

does the width. Of course there are observations where this relation does not hold, but the common trend for the whole dataset is somehow summarized in that red line. Some parts of the scatterplot fit more closely to the line than others, accounting for a smaller error to the prediction given by the red line. When representing data coming from different observations it is often important to include in our graph this uncertainty, the errors in the predictions, and measure the proximity to the prediction by a confidence interval (see Chap. 5 for more on linear models, prediction, error, and confidence intervals).

In order to show the errors and confidence intervals the ggplot2 library includes the geom_errorbars() function. These intervals are frequently depicted as a vertical bar constructed from the value of the observation plus or minus an error. The function geom_errorbars() features the arguments ymin and ymax for the lower and upper values, passed through aes().

Error bars can be used in addition to many different graphs such as histograms or bar plots, but a very common use is in time series. As an example, consider the economics dataset from ggplot2 library, a data frame of 6 economic variables measured monthly. As an example, we plot in Fig. 4.34 the monthly unemployment, measured in thousands of people, with an error bar of + one million people. Given the high number of observations in the time series above, the error bars (in green) appear as a wide region around the observations (in red).

```
min.unemp <- economics$unemploy - 1000
max.unemp <- economics$unemploy + 1000
ggplot(economics, aes(x=date, y=unemploy)) +
  geom_errorbar(aes(ymin=min.unemp, ymax=max.unemp), color=3) +
  geom_line(color=2, size=1.2)
```

When visualizing confidence intervals, one of the most useful and attractive tools available in ggplot2 is the geom_smooth() layer. Even before knowing the details of the mathematical methods, geom_smooth() visualizes these concepts without the need of explicit calculations to measure the error in the estimation.

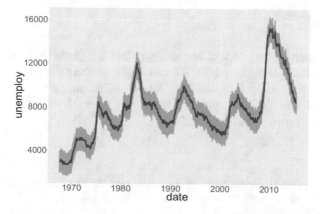

Fig. 4.34 Time series with error bars

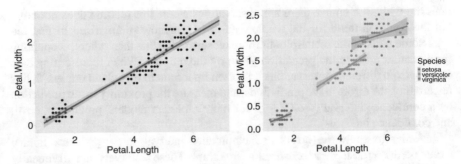

Fig. 4.35 Left: linear regression estimation. Right: several regressions fitted by `Species`

The main arguments, in addition to `mapping=aes()`, are `method`, that selects the method that is used to study the relation,[25] and `formula` which tells **R** what variables are to be related. Other arguments vary secondary features, such as `span`, that controls how adaptive the curves are, or `level`, that sets the level of confidence for the estimation intervals, by default `level=0.95`.

The following code produces a regression line with a 99.9% confidence interval, and the same regression but grouped by color, which produces not only the differentiation of the species but also each individual regression line. Both outputs are found in Fig. 4.35.

```
ggplot(iris, aes(x=Petal.Length, y=Petal.Width)) +
  geom_point() +
  geom_smooth(method=lm, level=0.999)

ggplot(iris, aes(x=Petal.Length, y=Petal.Width, color=Species)) +
  geom_point() +
  geom_smooth(method=lm, level=0.99)
```

4.2.3   Facets

As we have seen in the previous section, `ggplot2` is particularly efficient when working with categorical variables. We can easily produce plots where the data is colored depending on another variable. But we are missing a way of representing several graphics in the same plot, an equivalent to what we did with functions `par()` and `layout()` in Sect. 4.1.3.

---

[25]By default, a method is chosen based on the sample size. For less than 1000 observations, the method `loess` (*locally estimated scatterplot smoothing*, [5]) is implemented; otherwise, `gam` (generalized additive models, [8, 10]), is used. Both methods are far advanced for the scope of this book, the reason why we stick to the linear model by setting `method="lm."`

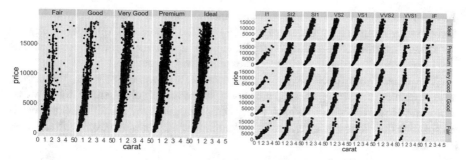

Fig. 4.36 Left: price and carat scatterplots classified by cut quality. Right: price and carat scatterplots classified by cut and clarity

With `ggplot2` not only can we reproduce the multi-plotting but the different plots to be produced can be somehow automated via *facets*. The main facet functions are `facet_grid()` and `facet_wrap()`, which are added preferably after the geometry layer. Let us explore the `diamonds` dataset to create several graphs featuring the same coordinates, but distinguishing different groups coming from a third variable, by means of `facet_grid()`.

```
ggplot(diamonds, aes(x=carat, y=price)) +
  geom_point() +
  facet_grid(. ~ cut)

ggplot(diamonds, aes(x=carat, y=price)) +
  geom_point() +
  facet_grid(cut ~ clarity)
```

Variables `carat` (a magnitude measuring diamonds weight) and `price` are used as X and Y axes of all plots, while `cut` (quality) and `clarity` are expressed as categorical variables, that we use to separate our sample and generate as many plots as categories they have. In the first call we only use the variable `cut` to generate columns via the code `facet_grid(.~cut)`.[26] When we make use of both variables in the second call, we have a matrix of 5 × 8 plots, each one corresponding to a different combination of values of `cut` and `clarity`.

The information that can be obtained from these scatterplots is broader. The relation between `price` and `carat` is growing (as expected) for all cut and clarity levels with a relation that appears to be nonlinear. However, this is not perfect and some scattering is found, specially for high prices and carats. In the first plot it can also be seen how high prices are less abundant for `Fair` diamonds, the lowest quality level. The second plot in Fig. 4.36 also brings some new details. We appreciate that `I1` diamonds with high prices are less common, and that they do need to have a considerable weight (`carat`) to reach the top `price`. Also, we

---

[26]To have the plot by rows we will use `facet_grid(cut~.)`.

have to mention how diamonds become increasingly rare when clarity is higher, being really strange for top clarity (IF) and the lowest cut quality (Fair).

Other arguments of facet_grid() are scales, space, as.table, switch, and drop, that are left to be explored in Exercise 4.11.

The other facet function is facet_wrap(). If we classify our dataset by one variable, facet_grid() creates a sequence of plots either by rows or columns. However, when we have many categories, we may certainly be interested in splitting the picture in several rows and columns and that is achieved with function facet_wrap().

```
ggplot(diamonds, aes(x=price, y=carat, color=color)) +
   geom_point() +
   facet_wrap(clarity ~ ., scales="free_y", ncol=4)
```

In Fig. 4.37 we reproduce the price vs. carat scatterplot segregated this time by clarity and colored by color (labeled from J—worst— to D—best—). Since we have 8 different clarity categories, we separate them into two rows with ncol=4. It is also important to highlight that for this final plot we are using four features, which implies that we are somehow creating a four variables graph, containing most of the conclusions previously obtained.

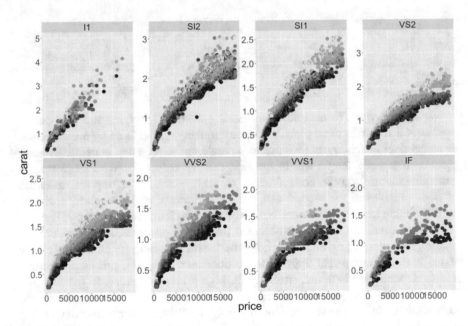

Fig. 4.37 Relation between the price and carat variable for 8 different clarity levels (plot colored with diamonds color)

4.2.4 Statistics

Some of the geometry functions from Sect. 4.2.2 include certain calculations which are performed automatically by calling them. Histograms and bar plots need to count the number of individuals per category before plotting them, while smooth functions estimate a model on data and then plot the resulting predictions. All these calculations are done internally by **R** using *statistical functions* that share the structure stat_command().

Most of them are masked inside the geometry functions, for example, geom_bar() uses stat_count() or geom_histogram() uses stat_bin(). Detailing and inspecting all these functions embedded in the ggplot2 commands will help to understand the statistics, beyond its geometric counterpart.

When studying a dataset, it can be useful to have a look to the *cumulative distribution* function (see Sect. 5.2) accounting for the cumulative frequencies of the sample, or the cumulative probabilities of a distribution in the theoretical setting. The interpretation of this function can be a bit harder than a histogram, but it is free of binning, and, therefore, we avoid rounding up our values into intervals. The bin length is a choice (made by the user or automatically) that may have an strong impact on the output, and using the cumulative distribution is a great alternative to avoid these situations. It can be performed using stat_ecdf(n=NULL) where n is the number of points where the cumulative distribution function is interpolated.

Using again the iris dataset we can easily plot the cumulative distribution function of the Sepal.Length, see the left panel in Fig. 4.38, with the code

```
ggplot(iris) + stat_ecdf(aes(x=Sepal.Length, color=Species))
```

which clearly shows that, for setosa flowers, lower lengths in sepals are more common than in the other two species. The curve increases slowly for values that are not observed so frequently and fast in ranges of high concentration. For instance, the cumulative distribution for the sepal length of setosa flowers grows quickly at first indicating that they are smaller, more precisely featuring an almost vertical increase

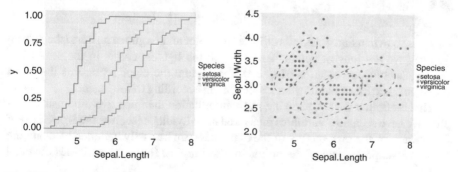

Fig. 4.38 Left: cumulative distribution of the iris dataset variable Sepal.Length. Right: elliptical confidence regions for the mean values of iris flowers

at around 5, where most lengths concentrate. The curves for versicolor and virginica follow a similar pattern but at bigger sepal length values.

Another interesting function is stat_ellipse(), used to implement *confidence regions* in our plot, defined in detail in Sect. 5.2. Arguments of this function are the type, which is the distribution used to compute the boundary region ("t," by default, for a t-student distribution, or "norm" for a normal distribution and "euclid" to draw a circle). Either way, ellipses or circles depend on radii set by level, indicating the confidence (default is 95%). For example, the right plot in Fig. 4.38 is produced with the code

```
ggplot(iris, aes(x=Sepal.Length, y=Sepal.Width, color=Species)) +
  geom_point() +
  stat_ellipse(type="norm", level=0.68, linetype=2)
```

This example shows that, using the normal distribution, the ellipses contain the 68% of the observations of each species. The overlapping ellipses of versicolor and virginica confirm our previous statement of mixed populations, while the setosa ellipse is clearly separated. Ellipses, as confidence regions, can be used to test hypothesis about two samples belonging to the same population.

The counterpart in ggplot2 for the function curve() that was studied in Sect. 4.1.1 is stat_function() that has the arguments as those in the base **R** graphics package. For example, Fig. 4.8 can be reproduced with

```
ggplot(data.frame(x=c(0, 2 * pi)), aes(x)) +
    stat_function(fun=sin, lwd=3, lty=2)
```

Another use of statistical functions recovers again the idea of visualizing more than two features in a two dimensional graph by means of color. We can *rasterize* a scatterplot with a surface function using stat_summary_2d(). This function automatically divides the plotting window into a grid of points using the arguments bins (number of grid cells) or binwidth (vector of vertical and horizontal width of each cell in the grid). Then, a function fun is evaluated at each location and plotted. As a function we can either use a third variable or a proper function (evaluated over this third variable in the points of the bin).[27] For example, the code

```
ggplot(iris, aes(x=Sepal.Length, y=Sepal.Width,
      z=Petal.Length)) +
  stat_summary_2d(fun="mean", binwidth=c(0.2, 0.2))
```

which is shown in Fig. 4.39, tells us how versicolor and virginica flowers (the bottom group) have the longest petals, associated to the lightest colors in the scale. The rectangular bins are squares of size 0.2×0.2 and the color represents the mean Petal.Length of the flowers inside the bin according to the legend.

This is another way of showing more information than just two dimensions in a plot. We have seen how variables measured on individuals (such as the Species of the flowers, or the cut quality of the diamonds) can be easily included by changing the color, shape, or size of the points in a scatterplot. However, variables obtained

---

[27]If an implemented function is used, it should be between commas.

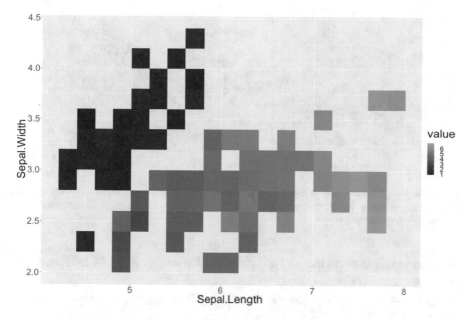

Fig. 4.39 Average petal length per sepal length and width

from additional calculations (such as the density of points or the variance of a third variable) cannot be showed this way directly. We require of functions like stat_density_2d() to create a new mapping and plot it over the scatterplot. Understanding the kind of variable we are using is fundamental when choosing which is the type of plot that better fits of our visualization purposes.

4.2.5 Customization Layers

The remaining element that can be combined with all the layers described in previous sections is a fine customization of certain additional features, such as coordinates, scale, transformations, labels, legends, or themes. These will finally help to present and communicate our analysis in a more professional way.

4.2.5.1 Coordinates

The last kind of layers that affect the way data is displayed are those for coordinates transformations. The first operation that can be performed is resizing axes by means of coord_cartesian() that shares with plot() the arguments xlim, ylim.

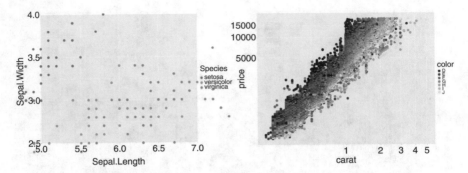

Fig. 4.40 Left: scatterplot with limited ranges on the axes. Points are allowed to appear outside the plotting window. Right: `carat` and `price` variables on the `diamonds` dataset after a logarithmic transformation

An easy example using the `iris` dataset can be seen in Fig. 4.40 left, produced with

```
ggplot(iris, aes(x=Sepal.Length, y=Sepal.Width, color=Species)) +
  geom_point() +
  coord_cartesian(xlim=c(5, 7), ylim=c(2.5, 4),
                  expand=FALSE, clip="off")
```

where `expand=FALSE` sticks to the limits set by `xlim` and `ylim` without allowing an additional margin, and `clip="off"` shows points outside plot margins, if any. Plot axes can be switched adding the layer `coord_flip()` with no arguments and the default ratio between y-axis and x-axis can also be specified using `coord_fixed(ratio)`.

4.2.5.2   Transformation of the Scale

Data analysis requires in many cases a mathematical transformation of the observed values. These transformations might be necessary after some theoretical approach or are simply useful to better visualize its content. They can be easily applied using `coord_trans(x, y)`. Any function can be used for transforming any of the variables but one of the most interesting ones is the logarithm ($f(x) = \log x$). The logarithm is defined as the inverse function of the exponential and, as such, transforms variables with an exponential growing behavior into linear plots that ease out the analysis and search of patterns.

Recall Figs. 4.36 and 4.37, where multiple plots of the `diamonds` dataset were depicted. They all consistently showed a direct relation between `price` and `carat` since, as one grows, the other does as well. This relation, though, does not seem to be linear as the `price` increases faster than `carat`. If we represent again the same data with a logarithmic transformation on its variables, we can see a clearer linear relation of the transformed variables (see Fig. 4.40 right).

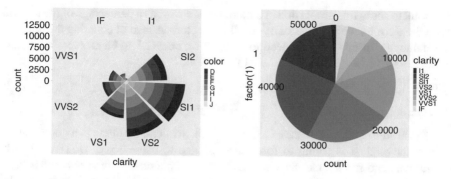

Fig. 4.41 Left: bar counts of diamonds clarity per color type. Right: colored bar counts of the diamonds clarity in polar coordinates

```
ggplot(diamonds, aes(x=carat, y=price, color=color)) +
  geom_point() +
  coord_trans(x="log", y="log")
```

It is important to learn how to read these plots in logarithmic scale, where the grid lines become closer as the variables get bigger. Logarithms expand the distances between small values and reduce those between big numbers. Hence, a small distance at the top and right of this plot is a much greater distance than the same variation at the bottom-left corner.

4.2.5.3 Polar Coordinates

Polar coordinates are a two dimensional system where points are defined by a distance from the origin r and an angle θ. Expressing variables in polar coordinates can be used to produce pie charts (recall function `pie()` from Sect. 4.1.2). As an example of this function we will adapt some bar plots and histograms. In Fig. 4.31 we used the `diamonds` dataset to make three plots. The second plot of Fig. 4.31 is a bar plot of `clarity` with variable `color` used to paint the bars. That same plot in polar coordinates is depicted in the left-hand side of Fig. 4.41 and is obtained with the code.

```
ggplot(diamonds, aes(x=clarity, fill=color)) +
  geom_bar() +
  coord_polar()
```

In this first approach, the values are depicted in the pie chart as "radial columns" of the same angle, whose radii are larger or smaller depending on the value. The variable to be displayed in the pie chart is passed with `theta`,[28] starting clockwise from the vertical (these values can be modified with `direction` and `start`). If

---

[28]The code above is equivalent to set `theta="x."`

we want to show this proportion in a more classic way, as a percentage of the area of the whole circle, we use `clarity` to fill the sectors instead of counting the number of diamonds. In this case we need to assign `factor(1)` to element x and define `theta="y"` in the `coord_polar()` function.

```
ggplot(diamonds, aes(x=factor(1), fill=clarity)) +
  geom_bar() +
  coord_polar(theta="y")
```

The need to add x=`factor(1)` might look artificial but there is a good reason for it: the analyst is trying to force the package to use `coord_polar()` with radial bars and not the classic areas. Even though pie charts are very popular they are misleading and usually present problems for human perception. It is frequently difficult to see what area is bigger, and color or order might deeply affect our perception of the plot. That is why pie charts should be used with extreme caution.

4.2.5.4   Labels, Titles, and Legend

By default `ggplot2` includes no main title and uses the names of the plotted variables as *labels* for the axes. If we want to override this information, we can do it with function `labs(title, subtitle, caption, tag, x, y)`. The following code will enhance all the archetypal `iris` plot with all possible labels (see Fig. 4.42).

```
ggplot(iris, aes(x=Sepal.Length, y=Sepal.Width, color=Species)) +
  geom_point() +
  labs(title="IRIS Sepals",
       subtitle="Setosa, Versicolor and Virginica",
       caption="Comparison between the sepal length and width
               for the iris dataset", tag="Fig. 1",
       x="Sepal Length", y="Sepal Width", color="Flowers")
```

Correctly naming a plot is key to an effective exposition. Labels make figures self-explanatory, adding any remark or contextual information necessary to

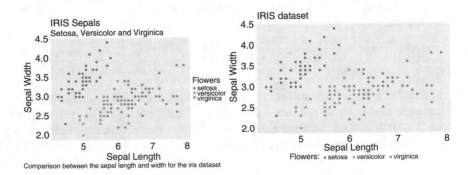

Fig. 4.42 Sepal length and width scatterplot with labels (left) and legend (right)

understand its content. Although unnecessary when exploring your data, always remember to include labels in a final plot.

Most of our plots include a *legend*, yet we never called a function that specifically creates or adjusts it. Legends are created by default when necessary, using the given variables and information, but if we want to modify this legend or simply remove it, we have to use the layer guides() and inside of it the argument color = guide_legend(title, title.position, labels, direction). With these elements we can change the appearance of the legend key, but moving it to another part of the plot has to be done through theme(legend.position). As an example of how to adjust the position and properties of a legend run the following code (see Fig. 4.42 right).

```
ggplot(iris, aes(x=Sepal.Length, y=Sepal.Width, color=Species)) +
    geom_point() +
    labs(x="Sepal Length", y="Sepal Width",
         title="IRIS dataset") +
    guides(color=guide_legend(title="Flowers:",
           title.position="left", labels=FALSE,
           direction="horizontal")) +
    theme(legend.position="bottom")
```

4.2.5.5 Saving a Plot

After producing an elaborated plot we may want to save it and publish it. Unlike graphics package, ggplot2 provides a unique saving function, ggsave(), with a similar usage to the saving functions from Sect. 4.1. The argument filename declares the name of the saved file and device selects the file format (such as "eps," "pdf," "jpeg," "png," "bmp," or "svg"). In addition, it accepts the path to the folder where the plot is to be saved, and the width and height of the saved plot in units ("in," "cm," and "mm"). By default, function ggsave() will save the last plot created in the current working directory, unless the argument plot is specified. This code serves as an example of this function.

```
ggsave("example_plot.png", width=14, height=11, units="cm",
       device="png")
```

4.2.6 Exercises

Exercise 4.9 Another **R** built-in dataset is LifeCycleSavings that contains interesting information on 50 different countries. The data was obtained by [13] and published by [2]. For a detailed description, access the help file by typing help(LifeCycleSavings). The following questions can be answered with help of different plots. Use the ggplot2 functions to visualize the information of interest.

(a) Compare variables pop15 and pop75. Can the population be classified into two or more different categories?
(b) Create a new variable in the LifeCycleSavings dataset with value "Old" when pop15 < 35 and "Young" otherwise.
(c) Repeat the plot from question a) with different colors for each category.
(d) Can we clearly separate variable pop75 with this classification? Create a frequency plot and check whether the resulting distributions overlap.
(e) Repeat the previous exercise with function geom_boxplot() and using variable dpi.
(f) Which of the three economical variables (sr, dpi, and ddpi) shows a stronger relation on the Old–Young category? Generate three different plots using pop15, pop75 and each one of the previous variables to visualize it.
(g) Plot again variables pop15 and pop75 with a linear regression estimation. Which is the expected value of pop75 for a country with pop15=37?

Exercise 4.10 The Theoph dataset contains the results of an experiment on the pharmacokinetics of theophylline over 12 patients. Reproduce the left image in Fig. 4.43. Use variables Time and conc to plot the evolution lines of the experiment for each patient in a single plot, with the following specifications.

1. Paint each line with a different color.
2. Add as well the average line in black.
3. Remove the legend.
4. Display labs and title as shown.

Exercise 4.11 Using the variables from the previous exercise, reproduce the right image in Fig. 4.43 by creating a multiplot with 12 individual evolution lines with the following details.

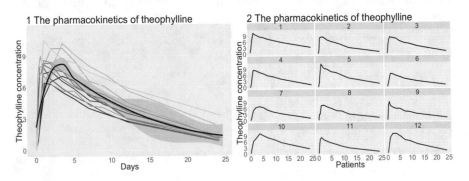

Fig. 4.43 Images for Exercises 4.10 and 4.11

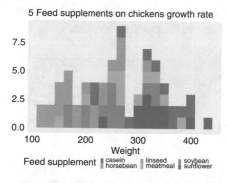

Fig. 4.44 Images for Exercises 4.12 and 4.13

1. Title the individual plots with the number corresponding to each patient.
2. Display 4 plots per column.
3. Display labs and title as shown.

Exercise 4.12 The `chickwts` dataset contains the results of an experiment conducted to measure the effectiveness of 6 different feed supplements on chickens. Build a histogram of the `weight` separated by `feed` type to represent the left image in Fig. 4.44 taking the following into consideration.

1. Use 20 bins.
2. Display labs and title as shown.
3. Move the legend to the bottom as shown.

Exercise 4.13 Continuing with Exercise 4.12, create now a pie chart of the chicken `weight` separated by `feed` type, so we can compare which food supplement has been used on more occasions during the experiment. Follow the instructions below to reproduce the right image in Fig. 4.44.

1. Display the `soybean` supplement at the top.
2. Display labs and title as shown.

4.3 Package `plotly`

In addition to the basic `graphics` and the more advanced `ggplot2`, **R** counts with many more visualization packages. This section deals with `plotly`, a set of tools for proficient and interactive visualization developed by Plotly Technologies Inc. [12] and later integrated in **R** by RStudio.[29] The outputs from `plotly` can be integrated in HTML codes for webpages or in specific applets. We will see some of

---

[29]`plotly` is also available in other languages such as Python and in standalone online versions.

the most useful and interesting tools, specially those that cannot be produced with
ggplot2 or that enhance the ones that we saw in the previous section.

To begin with, we load the library.

```
library(plotly)
```

4.3.1   Conversion of ggplot2 Plots

One of the most interesting resources of plotly is the conversion of ggplot2
plots to interactive plots: most of the images produced in the previous section
can be easily transformed into interactive plots with plotly, which enhances the
inspection possibilities and the beauty of the output. This conversion can be easily
done with function ggplotly().

```
iris.ggplot <- ggplot(iris, aes(x=Sepal.Length, y=Sepal.Width,
                                color=Species)) +
  geom_point() +
  guides(color=guide_legend(title="Flowers:"))
ggplotly(iris.ggplot)

iris.ggplot <- ggplot(iris, aes(x=Sepal.Length, y=Sepal.Width,
                                color=Species,
                                size=Petal.Width)) +
  geom_point()
ggplotly(iris.ggplot)
```

We only have to save the ggplot2 object and call ggplotly() to produce an
interactive image as seen in Fig. 4.45 left. Obviously, the interactive options cannot
be shown in a printed or in a .pdf version, but we present a screen capture when
hovering the cursor over an element of the plot: a label appears on the image. This
label contains a summary of all available information regarding this element, which
depends on the kind of plot and dataset. In this case the label shows the value of the
marked individual in other variables.

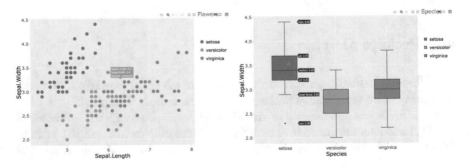

Fig. 4.45 Plots produced with ggplot2 and converted to interactive plotly images with
function ggplotly(). Left: iris dataset scatterplot. Right: iris dataset box plots

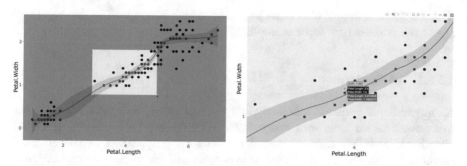

Fig. 4.46 Left: zooming on a ggplot2 converted to interactive plotly. Right: resulting image with multiple labels using button *Compare data on hover*

Depending on the chosen plot style, different labels appear. In Fig. 4.45 right we can see how in a box plot the highlighted information are the median and quartiles.

All plots generated by plotly feature a top bar of buttons that can be used to interact with the image. They include options such as download, zoom, or show multiple data labels. Zooming is specially interesting, and it can be done by simply clicking and dragging the mouse (see Fig. 4.46 left). A double-click will return the image to its original size. Button *Compare data on hover* (with a double label icon) displays as many labels as possible, enriching the plot with multiple information. An example can be seen in Fig. 4.46 right, after the previous zooming.

When panning we can click and drag the full image, moving towards other regions of the plot. Zooming and panning become of special interest when dealing with large amounts of observations, which cannot be easily distinguished in the original scale. All transformations can be reset with button *Reset axes* represented with a house icon.

4.3.2 Creating Plots with plotly

The plotly package, like ggplot2, is based on the Grammar of Graphics philosophy published by Wilkinson [17], however, the names and functions are entirely different. This package also allows to create highly advanced versions of most of the imaginable plot types, such as scatterplots, box plots, histograms, or surface densities. These are basically a repetition of what we learned in both previous sections with graphics and ggplot2, and, therefore, we do not describe in detail the usage of two dimensional representations with plotly. We will stick to the creation of 3D scatterplot with plotly, a functionality that was not available with ggplot2.

The main function in plotly is plot_ly(), which works similarly to ggplot(). The first argument of this function must be a data frame followed by the x and y variables. In addition, aesthetics can be introduced such as colors

or sizes. Then different traces can be added, which correspond to the `ggplot2`
layers. We will see now some examples.

4.3.2.1   3D Scatterplot

Adding multiple variables efficiently in a plot can greatly enhance its communicative power, as we have seen many times. 3D plots can be somehow difficult to visualize if the information is too scattered or overlapping. In fact, two observations might be far away but, when projected in two dimensions, might lay together generating a false perception of proximity. 3D scatterplots are usually not recommended as they can be deceiving unless, as it is the case now, the plot is interactive and we can freely rotate the image to visualize the data from any angle. In this situation, if the doubt arises of whether two points are close or not, a simple rotation will clarify the doubt.

After all previous sections we have a clear idea of the data content of `iris` but we can yet obtain a new vision with a 3D scatterplot as in Fig. 4.47.

```
plot_ly(iris, x=~Sepal.Length, y=~Petal.Width,
        z=~Sepal.Width) %>%
    add_markers(color=~Petal.Length, symbol=~Species,
                symbols=~c(15, 16, 18))
```

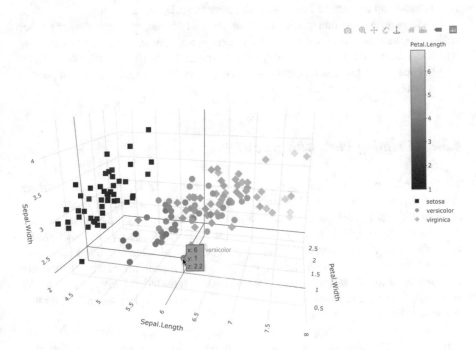

Fig. 4.47 3D scatterplot of `iris` dataset created with `plotly`. Image has been manually rotated for better visualization

The `plotly` grammar is remarkably similar to that of `ggplot2`. The data frame is the first argument followed by x, y, and z, the necessary arguments to build a 3D scatterplot, with the major difference being that the variables names must now be preceded by the $\sim$ symbol. New layers (now called *traces*) are added with `%>%` instead of the + symbol.

All traces in `plotly` share the structure `add_*()`, such as `add_markers()`, used for scatterplots. The arguments of these functions work like the aesthetics of `ggplot2`. Argument `color` can be used to paint the dots with a gradient of colors, using the fourth continuous variable and introducing then an extra dimension to the plot. In addition, with `symbol` and `symbols` arguments we tell the function to paint the dots with different shapes depending on the `Species` with shapes 15, 16, and 18. Finally, `plotly` conveniently adds grid lines when we move the mouse over the plot.

4.3.3 Exercises

Exercise 4.14 In Exercise 4.9 (c) we used the `LifeCycleSavings` data to obtain a scatterplot between variables `pop15` and `pop75`. We also added colors to distinguish between Young and Old countries, but now we want to see the name of the country when hovering over the markers. First, we need to create a new column with the countries names that we obtain from the row names:

```
LifeCycleSavings$Country <- rownames(LifeCycleSavings)
```

However, this column was not called in the original plot. Find out how to include it and convert to `plotly` to obtain Fig. 4.48 left.

Exercise 4.15 We now produce a `plotly` image without using a previously created a `ggplot2` object. As it can be seen in Fig. 4.48 right, we use again the

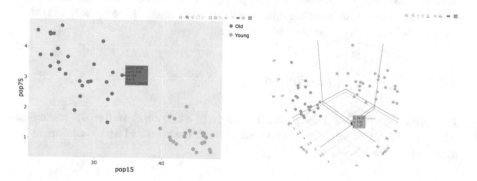

Fig. 4.48 Images for Exercises 4.14 and 4.15

LifeCycleSavings dataset to produce a 3D scatterplot. Try to reproduce this plot with the following specifications:

1. Use variables pop15, pop75, and sr, as axes x, y, and z.
2. Color the dots according to the Old–Young classification created in Exercise 4.10 (c).
3. Introduce the names of the countries using argument names.
4. Remove the legend with show.legend=FALSE.

4.4 Package leaflet

In the previous sections, we have seen how to create multiple types of plots and how to transform them into interactive images. Using well structured data frames we can build these plots from scratch, customizing every detail. However, data visualization also includes another way of working, where data is requested to huge online data sources (for example, using an API, as explained in Sect. 3.2.2) and processed locally in our computer.

We now devote a section to the leaflet library[30] [1, 4], a package developed by Vladimir Agafonkin[31] with the purpose of creating and customizing *interactive maps*.

Think of a *Google Maps* search. You are in a new city and you want to find in a map every restaurant in a neighborhood. Storing in your computer a detailed map of the whole planet is an unnecessarily high waste of memory. In addition, this information is constantly changing and we would need to update it every time we perform a new search. Also, the original search was about restaurants, but soon after we could be interested in pubs or in the need to find where hospitals are. Working with maps is a good example of this new use of data visualization, where we are constantly changing our data request on a big database.

We will see how to easily generate images of maps in **R**. On these maps we will be able to overplot icons, text, and polygons, marking locations and regions of our interest. In addition, some of the interactivity options that we have already seen with plotly will also be available with leaflet.

4.4.1 Maps and Coordinates

The main function of this package is leaflet() and it deeply resembles plot_ly(), as more content is added through the addTiles() command by means of the %>% operator.

---

[30]Leaflet is an open-source JavaScript library.

[31]https://agafonkin.com/.

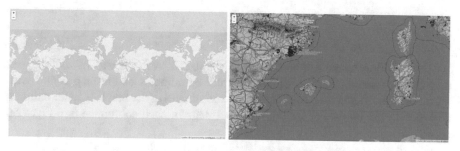

Fig. 4.49 World maps created with `leaflet`. Left: Default map created with OpenStreetMap. Right: Zoom of western Mediterranean Sea. © OpenStreetMap contributors

First, load the package.

```
library(leaflet)
```

Retrieving and plotting the world map to obtain the left-hand picture in Fig. 4.49 is as easy as

```
world.map <- leaflet() %>% addTiles()
world.map
```

As with `plotly`, we can zoom in and out and pan the map. This map is provided by OpenStreetMap [11], a free editable map of the world, that collects and produces a vast amount of data to be plotted. We zoom in over to see any city details in the map: streets, landforms, public transport lines, stores, etc.

Together with Open Street Map, `leaflet` provides more than 130 world maps from different sources (try typing `View(providers)` in the console to see a list of them). The way to visualize a different map from another source is by calling it with the function `addProviderTiles()`. For example, `MtbMap` (based on OpenStreetMap data) shows a physical map of Europe. Zooming in we can find Fig. 4.49 right.

```
leaflet() %>% addProviderTiles("MtbMap")
```

We can also produce maps of a specific region given by their coordinates with other methods of `leaflet`. The most basic one is `setView()` where we can specify the longitude and latitude coordinates of a certain location.[32] For example, the city of Los Angeles is located around latitude 39° N and longitude 118° W. We can particularize for a specific place in the city, as is the case of the Staples Center[33] with coordinates 34.04302° and −118.26725°, set with `lat` and `lng`, and zoom in with `zoom`.

---

[32]In geographic coordinates, latitude is the angular distance measured along a meridian, with value 0° at the equator and 90° at the north and south poles. Longitude is the angular distance from the Greenwich meridian along the equator going from 0° till 180° East and West, respectively. South latitudes and West longitudes will be set in **R** as negative.

[33]The Staples Center is a multi-purpose arena in Los Angeles city, site of several sports and arts international events.

```
leaflet() %>%
  addTiles() %>%
  setView(lat=34.04302, lng=-118.26725, zoom=16)
```

The resulting image can be seen in Fig. 4.50 left. Another way of zooming out a map is by setting interval bounds for the coordinates of the map center with fitBounds. In this code, we first set the coordinates of the Staples Center building and then fit the bounds of the map to ±0.1 degrees each coordinate (see Fig. 4.50 right).

```
LA.lat <- 34.04302
LA.lng <- -118.26725
leaflet() %>%
  addTiles() %>%
  fitBounds(lat1=LA.lat-0.1, lat2=LA.lat+0.1,
            lng1=LA.lng-0.1, lng2=LA.lng+0.1)
```

More data can be added to the map. The first step will be to mark the location of the Staples Center, which can be done with icons and text notes. With function addMarkers() we can create a location icon with a pop up message that appears when we click on it, as one can see in Fig. 4.51 left.

```
leaflet() %>%
  addTiles() %>%
```

Fig. 4.50 Left: Zoomed map centered at the Staples Center in Los Angeles. Right: Zoom out of the map with ±0.1° bounds. © OpenStreetMap contributors

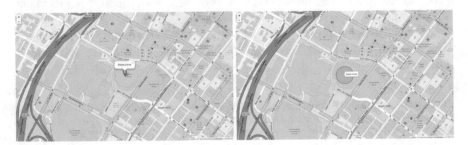

Fig. 4.51 Maps centered at the Staples Center in Los Angeles. Left: With a pop up message. Right: With a label and a 50 meters radius circle. © OpenStreetMap contributors

```
setView(lat=LA.lat, lng=LA.lng, zoom=16) %>%
addMarkers(lat=LA.lat, lng=LA.lng, popup="Staples Center")
```

Another alternative is to use labels, messages that appear when the mouse is over the icon. For example, we use the option `addCircles()` to create a circle of 50 meters radius around the Staples Center, as seen in Fig. 4.51 right. Arguments such as `color`, `stroke`, `weight`, and `fillOpacity` further customize the circle for better visualization.

```
leaflet() %>%
  addTiles() %>%
  setView(lat=LA.lat, lng=LA.lng, zoom=16) %>%
  addCircles(lat=LA.lat, lng=LA.lng, radius=50,
            color="green", stroke=TRUE, weight=5,
            fillOpacity=0.2, label="Staples Center")
```

4.4.2 Google Maps Platform

We have seen a few ways of plotting basic maps. But it is using larger datasets of locations when we can see the full potential of `leaflet`. Imagine you want to display a series of interesting places on a city map adding information related to these places. The user could manually provide all these coordinates and related information, one by one, but if the number of places is too high, we would rather prefer to help ourselves with an automatic tool. As seen in Sect. 3.2.2, APIs are interfaces that allow to obtain structured information from an online source. The most common API for geographical exploring is the Google Maps Platform,[34] a free limited service that offers enhanced service at different pricing. For a first approach, the free option is more than enough.

Log in is done at the cited page by clicking on the *Get started* bottom. We will select all data options (maps, routes, and places) and name our new project.[35] Once we have finished the process, the user's Google Map API key is provided. With this confidential code we can start downloading data from Google Maps.

First, we install and load the `ggmap` package, a library made to access the Google Maps API and download data. Then, we can register to Google Maps with function `register_google()` and our API key.[36]

```
library(ggmap)
register_google("w45I3C12h554i216N4s029m4E2i23K4I90N-356s")
```

This call produces no output but we can now start downloading data with function `geocode()`. As an example, we want to start by representing on a map 12

---

[34]https://cloud.google.com/maps-platform/.

[35]The usage of this platform requires to be registered and to accept Google terms and introduce our billing data, even though no charge is done without explicit user approval.

[36]This is a fake key used as an example, which should be substituted by the reader's personal one.

interesting places in the city of Los Angeles. First, we create a vector containing the names of the places, so we can order the search on Google Maps. The needed text is the same we would use on a regular Google search, i.e., a building name, an address, or a town, trying to specify enough details. If necessary, we can add extra information like the name of the city or country. We store the names of the selected places in variable

```
places.LA <- c("Staples Center", "Walt Disney Concert Hall",
               "University of Southern California",
               "Natural History Museum of Los Angeles",
               "Dodger Stadium", "Dolby Theather Hollywood",
               "Los Angeles Memorial Coliseum",
               "Los Angeles International Airport",
               "Dorothy Chandler Pavillion",
               "Cathedral of Our Lady of Los Angeles",
               "Los Angeles City Hall", "Griffith Observatory")
```

We can now order the search with `geocode()`, specifying the argument `output="more"` (for more information than just latitude and longitude) and the `source` where to retrieve the information from.

```
places.LA.goo <- geocode(places.LA, output="more",
                         source="google")
```

This creates `places.LA.goo`, a data frame[37] with longitudes, latitudes, and other relevant information. To this structure we attach the names of the places for further use, and plot a map with the places.

```
places.LA.goo$name <- places.LA
leaflet(places.LA.goo) %>%
  addTiles() %>%
  setView(lat=LA.lat, lng=LA.lng, zoom=11) %>%
  addMarkers(lng=~lon, lat=~lat, popup=~name)
```

To create the plot in Fig. 4.52, `data.LA.goo` is passed as an argument to function `leaflet()`. This allows an easier use of its content, calling the columns with the formula syntax (as in ~`lon` for the longitudes). As before, when clicking on an icon we will see the string created by `popup=~name`.

Another interesting function that can be used when dealing with multiple locations is `addPolygons()`. This function links points with a line connecting consecutive values. An area is created connecting the last point with the first. Selecting by the user some items of the previous list of places in Los Angeles, we can center the map in those places and delimit the area circumscribing the locations with a polygon (see Fig. 4.53).

```
center.places.LA <- places.LA[c(2, 9, 10, 11)]
center.places.LA.goo <- geocode(center.places.LA, output="more",
                                source="google")
```

---

[37]In fact it is a `tibble`: a data format used by **R** packages from the `tidyverse` universe like `leaflet`. In most situations it can be used just as a data frame.

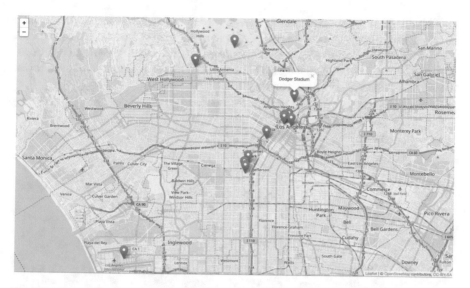

Fig. 4.52 Locations of 12 places of interest in Los Angeles. © OpenStreetMap contributors

Fig. 4.53 Polygon circumscribing a selection of locations in Los Angeles city center. © OpenStreetMap contributors

```
center.places.LA.goo$name <- center.places.LA
leaflet(center.places.LA.goo) %>%
  addTiles() %>%
  setView(lat=LA.lat+0.012, lng=LA.lng+0.02, zoom=15) %>%
  addPolygons(lng=~lon, lat=~lat)
```

We will yet provide another example, so we can have a better insight of the wide possibilities of this package. We will zoom out in our map of Los Angeles and create a map of the closer towns and neighborhoods to the city, showing their names, zone, and population. In addition, we provide a vector with the population of each town. We show the head of this dataset with code

```
name <- c("Los Angeles", "Glendale", "Thousand Oaks", "Calabasas",
          "Santa Monica", "Malibu", "Inglewood", "Burbank",
          "Pasadena", "Newport Beach",  "Santa Catalina Island",
          "Beverly Hills", "West Hollywood",
          "Palos Verdes States", "Lakewood", "Anaheim",
          "Fullerton", "Santa Ana", "Santa Clarita", "Venice")
population <- c(3792621, 203054, 128995, 24202, 92306, 12877,
               110598, 104834, 142647, 86160, 4096, 34484, 37080,
               13544, 154958, 352497, 140392, 334136,
               210888, 261905)
shire <- c("City", "North", "East", "East", "East", "East",
           "South", "North", "North", "West", "South", "East",
           "East", "South", "South", "West", "West", "West",
           "North", "East")
towns <- as.data.frame(cbind(name, population, shire))
head(towns)
```

```
          name population  shire
1    Los Angeles  3792621   City
2       Glendale   203054  North
3  Thousand Oaks   128995   East
4      Calabasas    24202   East
5   Santa Monica    92306   East
6         Malibu    12877   East
```

As before, we call the Google Maps API with function geocode() to obtain the coordinates and further information of each town.

```
towns.goo <- geocode(as.character(towns$name), output="more",
                     source="google")
```

Together with these locations, we would like to visualize every town name, the cardinal direction, and the population level. The names and populations will be included as popups and labels, but we will also draw circles around each town with sizes proportional to their populations. Since the population of the city of Los Angeles is much greater than any neighboring town, its circle will overlap the rest and we will use the opacity parameter, which works similarly to element alpha in ggplot2. First, we add new columns name and population to the towns.goo object. To convert strings into colors we create the column col with numbers 2 to 5 instead of the East, North, West, and South strings, plus assigning number 1 to the city of LA. Then, we use leaflet function colorNumeric() to assign colors to these values, creating a palette object; thanks to colors, the identification of each zone in the map will be much easier. The other new column, opa, will be used to control the color opacity of the drawn circles, with value 0.1 for Los Angeles city and 0.6 for the rest.

```
towns.goo$name <- name
towns.goo$population <- population
towns.goo$col <- c(1, 2, 3, 3, 3, 3, 4, 2, 2, 5, 4, 3, 3, 4, 4, 5,
                   5, 5, 2, 3)
pal <- colorNumeric(c("yellow", "red", "darkgreen", "blue",
                      "orange"), 1 : 5)
towns.goo$opa <- c(0.1, rep(0.6, 19))
```

Now we can create the final plot. Columns can be used with the function `addCircles()`. As we said, population is used to create the radii, dividing it by 50 to rescale. We show the final image in Fig. 4.54.

```
leaflet() %>%
  addTiles() %>%
  setView(lat=LA.lat-0.2, lng=LA.lng, zoom=8) %>%
  addCircles(lng=towns.goo$lon,
             lat=towns.goo$lat,
             radius=towns.goo$population / 50,
             color=pal(towns.goo$col),
             stroke=FALSE,
             fillOpacity=towns.goo$opa,
             popup=towns.goo$name,
             label=as.character(towns.goo$population))
```

With a few lines of code we are able to create a highly complex interactive plot showing multiple variables such as locations, names, or population levels. We can clearly see how the city of Los Angeles harbors much more that the entire population of the neighboring towns. A fast gaze on the plot also reveals that the population on the two towns on the East zone (Anaheim and Santa Ana) are considerably higher than the rest of towns in our dataset, for example.

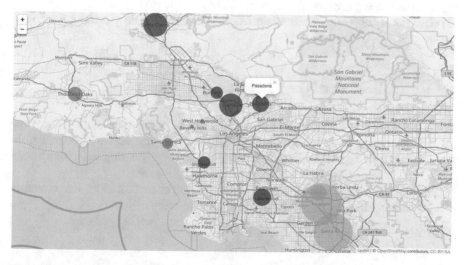

Fig. 4.54 Map the Greater Los Angeles area with town names and population levels. © Open-StreetMap contributors

This is just an example of what can be done with `leaflet` and the Google Maps Platform API. Today, many of the most important companies and institutions use `leaflet` to produce and exploit their geographical data, an immense contribution for data visualization.

4.4.3 Exercises

Exercise 4.16 One of the datasets provided by the default `graphics` package is `quakes`, with records of 1,000 earthquakes occurred in the Fiji Islands since 1964. We will analyze the distribution of these earthquakes with the help of the `leaflet` package. Let us begin by creating a map with the `quakes` dataset and the following specifications:

1. Find the coordinates of Suva, capital of the Fiji Islands and use it to center a map of longitude from $+13°$ to $-8°$ and latitude $\pm8°$. Use the map `OpenTopoMap` for a better terrain visualization.
2. Add an icon at the city of Suva with the popup message "Capital of the Fiji Islands."
3. Add the coordinates of the quakes as circles of radii 50 times its depth, black color, no stroke, and opacity 0.2. Include the magnitude of the quake as a label.

Now you should have an image like the one of Fig. 4.55. Use it to answer the following questions:

(a) How many fault planes of seismic activity do you find in your map?
(b) Which country is located to the East of the quakes main activity line?
(c) Do you think quakes tend to appear closer to the land?
(d) Deeper quakes happen near or far from the coast? Why do you think it is?

4.5 Visualization of the `flights` Dataset

We end the chapter doing several plots from the `flights` dataset to show how all the knowledge from this chapter fits into a real world scenario. Recall from Sect. 3.3.3 the `flights` database and assume that preprocessing has already been done.

We start by showing the altitude of the flight with code `IBE0730` that connects Madrid to Barcelona on a daily basis except for weekends.

```
flight <- flights[callsign == "IBE0730"]
plot(as.POSIXlt(flight$time, origin="2019-01-21"),
     flight$altitude.baro, type="o", pch=16, xlab="Time",
     ylab="Altitude", main="IBE0730 Flight Altitude")
```

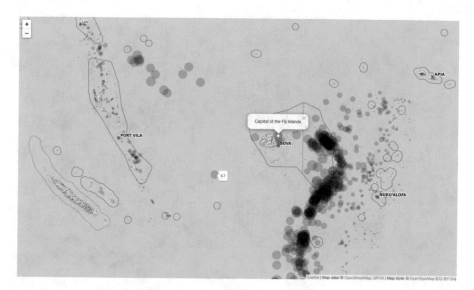

Fig. 4.55 Image for Exercise 4.16

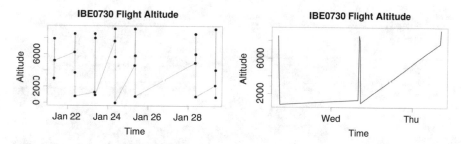

Fig. 4.56 Altitude of the flight IBE0730 over time

The image is in Fig. 4.56 left. Vertical lines look strange but it is very reasonable though. Recall that the resolution of the dataset is 15 min which means that a 1h flight (as the one we are considering) will be represented by at most four points, creating an almost vertical line (which seems totally vertical to the reader). The rest of the day, there is no flight with that code and there are no points. Notice as well that there are no flights on the weekend Jan 26th–27th 2019. The following code focuses just on Jan 22nd–24th, and is represented in the right picture of the same figure.

```
plot(as.POSIXlt(flight$time[5 : 13], origin="2019-01-21"),
    flight$altitude.baro[5 : 13], type="l", col=2, xlab="Time",
    ylab="Altitude", main="IBE0730 Flight Altitude")
```

The next code uses `ggplot2` to produce the following plots: Fig. 4.57 represents the frequency distribution of the altitude values, and the box plots of the distribution

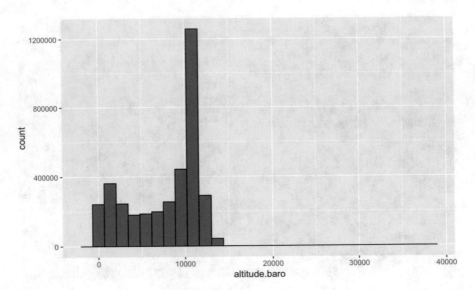

Fig. 4.57 Histogram of altitudes of the flights

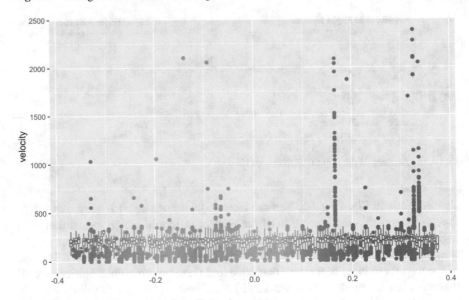

Fig. 4.58 Box plot of velocities of the flights, by country

of velocities are depicted by country in Fig. 4.58 (labels are removed due to the high amount of countries).

```
ggplot(flights) +
  geom_histogram(aes(x=altitude.baro),
  fill="DarkGreen", color="black")
ggplot(flights) +
```

```
geom_boxplot(aes(y=velocity, color=country)) +
theme(legend.position="none")
```

Note that, by far, the most frequent flight height is around 10,000 m. Regarding velocities, it looks like the distributions are similar regardless of the country, except for two countries, which happen to be Russia and United Kingdom (try ordering the flights by velocity and then do the box plot again removing these two countries.[38])

Another interesting visualization is that referring to spatial coordinates. By just calling

```
ggplot(flights) +
  geom_point(aes(x=longitude, y=latitude))
```

all points are plotted and, even with no map at the background, the shape of most continents in the World can be appreciated (see Fig. 4.58 left). It would be nice to use the leaflet package to plot the positions of the planes over a map with a code such as

```
leaflet() %>%
 addTiles() %>%
 addMarkers(lng=flights$longitude, lat=flights$latitude)
```

The problem, though, is that plotting almost four million point is a hard task and in fact the ggplot() above surely took some time to finish. If those nearly four million points are interactive objects, the capabilities of any personal computer are clearly exceeded (the code above can be tried safely for 10,000 points). For a simple map representation we can use the library rworldmap together with the basic plot() function as follows

```
library(rworldmap)
newmap <- getMap()
plot(newmap)
points(flights$longitude, flights$latitude, col="blue", pch=16,
       cex=.6)
```

to obtain the right-hand picture in Fig. 4.59. At a glance, interesting economic conclusions can be extracted if the amount of flights are interpreted as a measure for development and wealth of a country or region; Africa is almost empty, while Europe or North America is much more crowded.

---

[38]Try   flights[!(country=="Russian Federation" | country=="United Kingdom")].

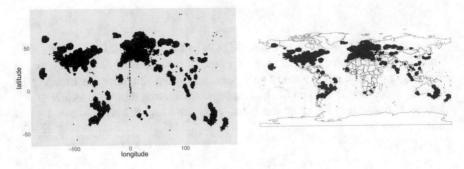

Fig. 4.59 Plot of latitude and longitude of all flights, and the same plot over a World Map

References

1. Agafonkin, V. Leaflet: an open-source JavaScript library for mobile-friendly interactive maps. https://leafletjs.com/, 2014. [Online, accessed 2020-02-29].
2. Belsley, D.A., Kuh, E. and Welsch, R.E. Regression diagnostics: Identifying influential data and sources of collinearity. *Wiley Series in Probability and Mathematical Statistics*, 571(4), 1980.
3. Chambers, J. M., Cleveland, W. S., Kleiner, B. and Tukey, P. A. *Graphical Methods for Data Analysis*. Wadsworth & Brooks/Cole Mathematics Series, Springer, Heidelberg, Germany, 1983.
4. Cheng, J., Xie, Y., Wickham, H. and Agafonkin, V. leaflet: Create interactive web maps with the JavaScript 'leaflet' library. *R package version*, 1(0):423, 2017.
5. Cleveland, W.S. LOWESS: A program for smoothing scatterplots by robust locally weighted regression. *American Statistician*, 35(1):54, 1981.
6. Cleveland, W.S. *The Elements of Graphing Data*. Wadsworth Publ. Co., California, USA, 1985.
7. Fisher, R.A. The use of multiple measurements in taxonomic problems. *Annals of Eugenics*, 7(2):179–188, 1936.
8. Hastie, T., Tibshirani, R. and Friedman, J.H. *The elements of statistical learning: data mining, inference, and prediction*. Springer, Berlin, Germany, 2009.
9. Holtz, Y. Margin and Oma cheatsheet. https://www.r-graph-gallery.com/74-margin-and-oma-cheatsheet/, 2016. [Online, accessed 2020-02-29].
10. James, G., Witten, D., Hastie, T. and Tibshirani, R. *An Introduction to Statistical Learning: With Applications in R*. Springer Publishing Company, Incorporated, New York, USA, 2014.
11. OpenStreetMap. Planet dump retrieved from https://planet.osm.org. https://www.openstreetmap.org, 2017. [Online, accessed 2020-02-29].
12. Plotly Technologies Inc. Collaborative data science. https://plot.ly, 2015. [Online, accessed 2020-02-29].
13. Sterling, A. Unpublished BS Thesis, 1977.
14. H. et al. Wickham. Welcome to the tidyverse. *Journal of Open Source Software*, 4(43):1686, 2019.
15. Wickham, H. ggplot2: Elegant graphics for data analysis. https://ggplot2.tidyverse.org, 2016. [Online, accessed 2020-02-29].
16. Wickham, H. and Grolemund, G. *R for Data Science: Import, Tidy, Transform, Visualize, and Model Data*. O'Reilly Media, Inc., California, USA, 2017.
17. Wilkinson, L. *The grammar of graphics*. Springer Science & Business Media, Berlin, Germany, 2006.

Chapter 5
Data Analysis with R

Data is all around us: sensors, messages, chips, cellphones, webs, measurements...The beyond exponential growth of data and its complexity is making necessary to use, adapt, create and improvise new methods of data analysis, sometimes making use of more advanced mathematics and statistics, new and faster algorithms and methodologies being capable of analyzing the vast amount of data available.

The data analysts of today should have a strong and wide knowledge about the statistical foundations and developments of data analysis in order to deal with the new methods and tendencies that are being used and the ones that are yet to come. From the very first ideas about probability from Pascal, Laplace, Fermat, or Bernoulli, through the development of modern statistics by Pearson and Fisher, the *ordinary least squares* by Gauss or the *Bayesian statistics*, until the *machine learning* algorithms and *deep statistical learning* methods that are shaping the way we understand our world, everything is about analyze and understand data.

In this chapter we introduce the statistical basis for the data science in a simple and somehow algorithmic way, without mathematical proofs but providing motivation and justification for the procedures explained. Although we start from the very beginnings, the material in this chapter (together with the rest of the book) should provide enough insight and techniques to perform successful analysis on possibly very large datasets.

We begin by the introduction of the *descriptive statistics*, studying the statistical characteristics and measures of a single variable. Fundamental concepts as the mean, median, quartiles, variance, standard deviation, skewness, and kurtosis among others are introduced. After that, we present the basics of *probability theory* as the main tool for *statistical inference*, where we use probability theory to generalize and estimate values about a huge set of individuals using measures taken in smaller samples. Once is clear how to analyze a single variable, we introduce the study of relationships among them, using the *linear correlation coefficient* and the *linear regression*. We also study how to relate categorical variables, which are

© Springer Nature Switzerland AG 2020
A. Zamora Saiz et al., *An Introduction to Data Analysis in R*, Use R!,
https://doi.org/10.1007/978-3-030-48997-7_5

very common and demand special attention. We finish the chapter by introducing the *multiple linear regression* and the *logistic regression*, to model and forecast numerical and categorical variables using other variables as predictors.

The concepts in the chapter are introduced using a dataset as reference, showing the necessary **R** code to perform the analysis. More examples with different properties are included for the sake of completeness.

5.1 Descriptive Statistics

Since data collection and preprocessing were studied in Chap. 3, we assume for the whole chapter that we have performed all the necessary refinement of our data, as seen in Sect. 3.3, and that our dataset is ready to be investigated. We make use of the datasets embedded by default in **R**, as well as other datasets already prepared for examples such as the ones we have been working with along the book.

We start off analyzing the Pima Indian dataset for diabetes from the UCI Machine Learning Repository. It is a dataset of 9 physiological measures, such as glucose, mass index, and blood pressure of 768 women of the Pima tribe located in south Arizona.[1] We load the dataset `PimaIndiansDiabetes2`.[2] As explained in Sect. 3.2.1, this set is provided within the `mlbench` library so we load it with `data()` and do a first inspection with the **R** help and `head()`.

```
library(mlbench)
data(PimaIndiansDiabetes2)
?PimaIndiansDiabetes2
head(PimaIndiansDiabetes2)
```

	pregnant	glucose	pressure	triceps	insulin	mass	pedigree	age
1	6	148	72	35	NA	33.6	0.627	50
2	1	85	66	29	NA	26.6	0.351	31
3	8	183	64	NA	NA	23.3	0.672	32
4	1	89	66	23	94	28.1	0.167	21
5	0	137	40	35	168	43.1	2.288	33
6	5	116	74	NA	NA	25.6	0.201	30

```
  diabetes
       pos
       neg
       pos
       neg
       pos
       neg
```

It is formatted as a data frame of 9 variables (8 numerical and one factor: `diabetes`) with 768 observations. The final goal is to extract joint information

---

[1] See https://archive.ics.uci.edu/ml/datasets/diabetes from the UCI Machine Learning Repository.

[2] This dataset is a corrected version of the original, `PimaIndiansDiabetes`, where values which are impossible in real life (such as a zero value for glucose) are replaced by NAs.

about the dataset, that is, we want to know how the variables are related to each other and how one of the variables is explained (if possible) using the rest. In this case, we would like to know how these physiological measures explain or affect the others and how we can determine whether a woman is likely to present diabetes or not, by using the variables in the table. For example, are the physiological measures greater or smaller than average in women presenting diabetes? The study of relations among several variables is called *multivariate statistics* and we shall present it later. Before that, we consider every variable independently and perform an analysis of each of them without considering their interactions. The analysis of a single variable is called *univariate statistics*.

We begin analyzing single variables with some descriptive analysis. Each variable is a set of numbers and we sometimes refer to it as an *univariate distribution*. The first step is to describe the distribution of each variable with some measures designed to answer specific topics. For example, what is the average number of times that a woman in the dataset has been pregnant? What is the number that best represents the pressure for women in these data? Is there a wide variety of data values (dispersion) or, on the contrary, do all numbers range in a small set? Are there any extreme values? *Descriptive statistics* is about computing quantities that allow to understand the data and summarize it. The idea is to obtain a set of descriptors to encode all possible information. With these descriptors, data features should be explained. We introduce four types of measures: position, dispersion, shape, and concentration in this section. The functions to implement these functions, unless otherwise stated, belong to default package stats.

If we want to call the variables of a dataset directly, as vectors, without the need of using the $ syntax (as shown in Sect. 2.2.5), we can use the attach() function. We assume throughout this section that the dataset PimaIndiansDiabetes2 is attached.

```
attach(PimaIndiansDiabetes2)
```

```
The following object is masked from package:datasets:pressure
```

The warning message means that there is another object called pressure in a loaded library (indeed, in datasets) so, in order to avoid conflicts, our call will prevail and the other object will be ignored.

5.1.1 Position Measures

Quantitative variables take values in a certain set, which can be either a discrete number of values or all real numbers, for example. We want to describe this whole set of values by certain quantities summarizing it. This way, instead of considering all different values, we reduce this information to position measures, indicating where the distribution of values is located.

5.1.1.1 Central Measures

One of the first questions to answer when we have a list of numerical values is whether a notion of center exists and where to find it. But what is actually the meaning of a center? We want a number representing all the values at once and the best candidate is the one which is the closest possible to all of them. This is achieved by the *arithmetic mean*, also called the *average*, which is the sum of all the numbers divided by the total number of values. Therefore, the arithmetic mean of the n numbers $X = \{x_1, x_2, \ldots, x_n\}$ is computed by the formula

$$\overline{x} = \frac{\sum_{i=1}^{n} x_i}{n}.$$

The **R** function for the average is mean(). As an example, the numbers

$$4, 5, 8, 2, 8, 4, 5, 9, 2, 7$$

could represent the grades of ten students in a subject with a grading system that ranges from 0 to 10 and we can compute the mean

```
mean(c(4, 5, 8, 2, 8, 4, 5, 9, 2, 7))
```

```
[1] 5.4
```

meaning that the average grade of the students is 5.4. The question of whether this average is representative as an overall grade for the ten students or not will be addressed in Sect. 5.1.2, where the concept of dispersion is introduced.

For the Pima tribe dataset, we can compute the mean of some of the variables, such as mass.

```
mean(mass, na.rm=TRUE)
```

```
[1] 32.45746
```

We need to set the na.rm=TRUE in order to ignore the possible NAs since, even when our dataset is free of inaccuracies, it may contain some (if not many) missing values that should be removed for a proper calculation of the mean. We can also easily obtain the means of all variables using function sapply.

```
sapply(PimaIndiansDiabetes2[, -9], mean, na.rm=TRUE)
```

```
  pregnant      glucose     pressure      triceps
 3.8450521  121.6867628   72.4051842   29.1534196
   insulin         mass     pedigree          age
155.5482234   32.4574637    0.4718763   33.2408854
```

Notice how we remove column 9, corresponding to the categorical variable diabetes, from our calculation. This is one of the first and most elementary insights we can obtain from a dataset, and it tells us things such as that in a sample of women with an average age of 33 years, the mean of times each one has been pregnant is close to 4. If this is not the first time we analyze a dataset of this kind, we can compare our results with our previous knowledge, hence obtaining a first image of which kind of data are we working with.

The arithmetic mean is the most popular and widely used measure for computing the average of a list of numbers. There are some situations though, when the arithmetic mean does not correspond to the intuitive idea of an average that we have in mind and we need to use alternative statistics. This will always depend on the kind of data we are using, data of a different nature than the one used in our diabetes sample.

In order to illustrate these cases, we propose the following toy example: suppose we have collected the percentage evolution for a shop sales during 5 years, obtaining the following rates of change: 10%, −5%, 20%, 5%, and −3%. That means that if the sales at the beginning were 5,000 euros, then after the first year, the sales are a 10% higher, this is 5,000 multiplied by 1.10, 5,500 euros. The second year, the sales are 5,000 times 1.10 times 0.95, making 5,225, and so on. Hence, after 5 years the sales are the product of 5,000 successively incremented by each factor, that is,

```
sales.per <- c(0.10, -0.05, 0.20, 0.05, -0.03)
sales <- 1 + sales.per
5000 * prod(sales)
```

```
[1] 6385.995
```

On the other hand, the concept that the average should represent in this case is a percentage such that, if the evolution of sales were constant and equal to that average during the 5 years, then the final result would be the same:[3]

$$6,385.995 = 5,000 \cdot (1+G)^5 \longrightarrow G = \sqrt[5]{\frac{6,385.995}{5,000}} - 1 = 0.050151.$$

Therefore, an approximate annual increment of a 5% in our sales would produce, after 5 years, an equivalent outcome to the previous rates.

```
5000 * (1.050151) ^ 5
```

```
[1] 6385.998
```

As a matter of comparison, we can use the arithmetic mean $\bar{x} = 0.054$, obtaining a different result:

```
5000 * (1.054) ^ 5
```

```
[1] 6503.888
```

---

[3]This is the standard concept of the mean: it is the same to pay 1€, 2€ , and 3€ , respectively, for three products than paying the average price (2€ per unit) for each of them.

It is a common mistake to use the arithmetic mean when it is not suited. A cumulative growth, such as rates of change and other time evolved quantities, is made of ordered values, where its position is necessary to understand it. When this is the case we should consider using the *geometric mean* instead. For a sample X of n values $x_1, x_2, \ldots, x_n$, the geometric mean is the nth-root of its product, that is,

$$G = \sqrt[n]{x_1 \cdot x_2 \cdot \ldots \cdot x_n}.$$

R has no built-in function to calculate the geometric mean but several packages include their own versions. We can load the library `pysch` and use function `geometric.mean()`. However, this function makes use of the logarithmic function and will show a warning message when negative values are found. To avoid this problem we can use the incremental values $1 + x_i$:

```
library(psych)
geometric.mean(sales) - 1
```

```
0.05015091
```

Statisticians still consider other ways of calculating the mean depending on the nature of the data. As we can see, understanding the origin and meaning of data is fundamental to correctly choose the appropriate statistic.

One of these situations happens when we deal with rates per unit or speeds. Here, suppose we have been riding our bike uphill the Tourmalet pass in France along its 23 km with an average speed of 10 km per hour. Exhausted, at the top, we come back and go downhill the same 23 km with a speed of 30 km per hour. What is the average speed in the total 46 km? A naive (and wrong) approach would be to use the arithmetic mean to conclude that the average is 20 km per hour, which does not correspond to reality. If we compute the total elapsed time in such a ride, we have

```
23 / 10 + 23 / 30
```

```
[1] 3.066667
```

But, considering 20 as the average speed, we would obtain a different number.

```
46 / 20
```

```
[1] 2.3
```

What is going on in here? Although we ride the same distance (23 km), much more time is spent while going uphill and therefore the speeds should be averaged in different proportions. Hence, we need a new measure for averaging these kinds of scenarios. We define the *harmonic mean* of n values $X = \{x_1, x_2, \ldots, x_n\}$ as the number

$$H = \frac{n}{\frac{1}{x_1} + \frac{1}{x_2} + \ldots + \frac{1}{x_n}}.$$

The function `harmonic.mean` from the `psych` package computes this in **R**.

```
speeds = c(10, 30)
harmonic.mean(speeds, na.rm=TRUE)
```

```
[1] 15
```

Now, if we compute the time riding with this new average speed of 15 km per hour, we obtain the expected result.

```
46 / 15
```

```
[1] 3.066667
```

Generally speaking, for speed or rates per unit of time or distance, and other variables defined as quotients, the harmonic mean is the appropriate average to use.

Beyond the different means we have seen so far, there are other measures to define or compute the center of a set of numbers. The *median* of n numbers $X = \{x_1, x_2, \ldots, x_n\}$ is the value, usually denoted by Me or M_e, such that half the numbers are less than it and the other half are greater. If the number of elements n is odd, then there is a value in the set that lays exactly in the middle, hence it is the median. If the number of elements is even, the median is defined as the arithmetic mean of the two central values, when ordered. The function `median()` computes the median in **R**.

```
median(1 : 7)
```

```
[1] 4
```

```
median(1 : 6)
```

```
[1] 3.5
```

In our diabetes dataset, the median for `mass` is not very different from its mean.

```
median(mass, na.rm=TRUE)
```

```
[1] 32.3
```

For symmetric variables, both values will be the same, although the opposite is not necessarily true. We will study the symmetry or skewness of a variable in greater detail in Sect. 5.1.3.

The median has an important property, which is being resistant to extreme values. That is, if we add another number (or a few) which is much bigger or smaller than the rest, the mean will shift noticeable. However, the median is almost unaffected by this phenomenon. Using the previous example,

```
mean(c(1 : 7, 30))
```

```
[1] 7.25
```

```
median(c(1 : 7, 30))
```

```
[1] 4.5
```

Introducing a highly different value, such as 30, we can see how the mean is shifted from 4 to 7.25, which is only smaller than the biggest number. The median instead was only increased from 4 to 4.5, halfway distance to the next value in size.

A single value is strongly affecting the value of a measure for the whole sample and that might be a problematic outcome. This kind of highly different values are called *outliers* and need of special care in data analysis, so they will be introduced later in detail. The mean is very sensitive to outliers and thus not recommended in presence of them or, at least, recommended only if carefully used. The median, however, is resistant to the presence of outliers, and can be very useful when having abnormal values.

Now we can compare the means obtained for `PimaIndiansDiabetes2` dataset with their medians:

```
sapply(PimaIndiansDiabetes2[, -9], median, na.rm=TRUE)
```

```
pregnant   glucose  pressure    triceps
  3.0000  117.0000   72.0000    29.0000
 insulin      mass  pedigree        age
125.0000   32.3000    0.3725    29.0000
```

Some variables, like `pressure`, `triceps`, and `mass`, show very similar values. Others, like `pregnant`, `insulin`, and `age`, are strongly changed, indicating a lack of symmetry on these variables. This conclusion already indicates us that these variables would probably require of a deeper analysis to fully understand them.

There is one more central measure we mention here. The *mode* of n numbers $X = \{x_1, x_2, \ldots, x_n\}$ is the most frequent value, usually denoted by Mo or M_o. If we understand our sample as the realization of a random variable (see Sect. 5.2), the mode is the value with a higher probability of being observed. This is clearly a representative value when we are interested in predictions. Note that there might be just one mode (unimodal population) or there could be more than one mode if there are two (or more) values which happen to be the most frequent. Again, **R** does not provide a built-in function to calculate the mode. Even if it is not complicated to write our own function, we recommend one of the functions provided by the **R** packages, such as `DescTools`. For the variable `mass` in our diabetes dataset we obtain its mode with function `Mode()` [4]

```
library(DescTools)
Mode(mass)
```

```
[1] 32
```

which, as we can see, is the same value obtained for the median. One elemental property of the mode and the median (of an odd number of values) is that the result is always an element of the sample. For the arithmetic mean (and also the geometric and harmonic means) this is not true in general.

---

[4] Remember to type the `Mode()` function with uppercase M, function `mode()` instead returns the class of an object.

Again, we can obtain the mode of all variables with

```
sapply(PimaIndiansDiabetes2, Mode, na.rm=TRUE)

$pregnant
[1] 1

$glucose
[1]   99 100

$pressure
[1] 70

$triceps
[1] 32

$insulin
[1] 105

$mass
[1] 32

$pedigree
[1]  0.254 0.258

$age
[1] 22

$diabetes
[1] "neg"
```

The output appears somehow different this time. As mentioned, the mode of a sample is not necessarily unique, and when two or more elements are found to be the most frequent they are returned as a vector: this is the case of glucose and pedigree. That forces the function to coerce the output object into a list containing both numeric and vector elements. In addition, the mode function can also be calculated on categorical variables, such as diabetes, as opposed to the previous measures. We see now that most of the individuals in the dataset do not suffer from diabetes. If we compare the mode value with the means and medians obtained previously we will see again differences. Variables with differences between mean and median, like pregnant, insulin, and age, show again different values for the mode, reinforcing our initial conclusion of a lack of symmetry. Others, like the variable mass, have a mode similar to its mean and median, a necessary, yet not sufficient, condition for symmetry.

As we have seen, the center of a set of numbers can be computed in several different ways, depending on the needs of each case. We can also use the visualization tools for univariate distributions studied in Sect. 4.1.2.[5] In Fig. 5.1,

---

[5]We use the command abline() as well, which is the equivalent inside the base graphics package to geom_line() studied in Sect. 4.2.2.

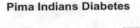

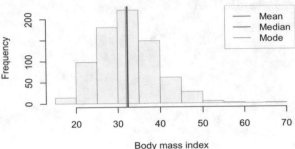

Fig. 5.1 Histogram of variable mass

the histogram of the mass variable of the Pima Indian Diabetes dataset together with the three central measures that we just explained are depicted

```
# plot histogram
hist(mass, col="seashell", border="black",
    xlab="Body mass index",
    main="Pima Indians Diabetes")
# add mean
abline(v=mean(mass, na.rm=TRUE), col="red", lwd=2)
# add median
abline(v=median(mass, na.rm=TRUE), col="blue",  lwd=2)
abline(v=Mode(mass), col="seagreen", lwd=2)   # add mode
legend(x="topright", c("Mean", "Median", "Mode"),
       col=c("red", "blue", "seagreen"), lwd=c(2, 2))  # legend
```

The relative positions of the central measures are an important characteristic of a distribution. We will go deeper into this topic in Sect. 5.1.3.

5.1.1.2 Quantiles

Once we have located the center of the distribution (for different definitions of center), we may want to map the distribution of the values with further details. A central position is far from being enough to fully characterize a sample of values and we need to locate more informative positions.

As a generalization of the median, we can compute the values that partition the distribution into subsets of a given percentage of the values. The *p quantile*, where p is a number between 0 and 100, of n numbers $X = \{x_1, x_2, \ldots, x_n\}$ is the value Q_p which is greater than $p\%$ the data. For example, in the mass variable, the 65 quantile is computed with function `quantile()`:

```
quantile(mass, probs=0.65, na.rm=TRUE)
```

```
 65%
34.54
```

We knew from the median calculation that value 32.3 leaves 50% of values below it. If we increase that value to 34.54, we will be leaving a 65%. The most important quantiles are $Q_{25}, Q_{50}, Q_{75}, Q_{100}$ which are usually referred to as *quartiles* and denoted by $Q1, Q2, Q3$, and $Q4$, respectively. Note that the second quartile $Q2$ is exactly the median Me and the fourth quartile $Q4$ is the maximum value. The `quantile` function computes the quartiles if no number is given to the `probs` argument.

```
quantile(mass, na.rm=TRUE)
```

```
  0%   25%   50%   75%  100%
18.2  27.5  32.3  36.6  67.1
```

The quartiles divide the distribution into four equal parts and this is a very common measure to use in any descriptive analysis. Another popular measure are the *deciles* which are the values breaking up the distribution into laps of a 10%. We compute the deciles for the `mass` variable.

```
quantile(mass, probs=seq(0, 1, 0.10), na.rm=TRUE)
```

```
   0%    10%    20%    30%    40%    50%    60%    70%    80%    90%   100%
18.20  24.00  26.20  28.40  30.34  32.30  33.80  35.50  37.88  41.62  67.10
```

Looking at the obtained values we can see how the first values rapidly increase from the minimum, while after the second decile (20%), the next 10% of values is contained in an interval of around 2 units. The last two deciles (90% and 100%) show instead much larger values, indicating a long tail to the right. These conclusions can be seen as well from the histogram in Fig. 5.1, with bars growing rapidly before 30, where they start a smooth evolution. The large atypical values on the right of the value 50 are the long right tail. As we can see, variable `mass` is an example of a variable of nearly coincident central positions with non-symmetry.

The inverse operation can be of interest as well: given a certain quantity we want to know to which quantile corresponds, i.e., which is the percentage of values below the quantity. In **R**, we can calculate it by using function `ecdf()`.[6] Given a data sample, such as the variable `mass`, `ecdf(mass)` creates a new function, which we can use to obtain the proper percentages. For example, for the value $x = 40$, we have

```
percent.fun <- ecdf(mass)
percent.fun(40)
```

```
[1] 0.8731836
```

and, at the median, we obtain (approximately) the expected 50%:

```
percent.fun(median(mass, na.rm=TRUE))
```

```
[1] 0.5019815
```

---

[6]The name of the function `ecdf` is the acronym of Empirical Cumulative Distribution Function, which will be explained in Sect. 5.2.

As we have mentioned, the quantiles at 0% and 100% ($Q4$) correspond to the minimum and maximum values of the sample, which can be obtained with functions min() and max(), respectively (see Sect. 2.2.2).

The quantities described in this section are fundamental measures of any random sample and provide a fast and informative insight. Given their importance, **R** provides the function summary(), already used in Sect. 2.2.4 for categorical variables. On a numerical variable, it returns the minimum and maximum values, the 1st and 3rd quartile, the mean and the median:

```
summary(mass)
```

```
   Min. 1st Qu.  Median    Mean 3rd Qu.    Max.    NA's
  18.20   27.50   32.30   32.46   36.60   67.10      11
```

Note that now we do not include the flag na.rm=TRUE; nevertheless the function counts the number of NAs. In addition, it can be used directly on a full dataset, obtaining the positions for all variables

```
summary(PimaIndiansDiabetes2)
```

```
   pregnant           glucose            pressure          triceps
 Min.   : 0.000   Min.   : 44.0    Min.   : 24.00    Min.   : 7.00
 1st Qu.: 1.000   1st Qu.: 99.0    1st Qu.: 64.00    1st Qu.:22.00
 Median : 3.000   Median :117.0    Median : 72.00    Median :29.00
 Mean   : 3.845   Mean   :121.7    Mean   : 72.41    Mean   :29.15
 3rd Qu.: 6.000   3rd Qu.:141.0    3rd Qu.: 80.00    3rd Qu.:36.00
 Max.   :17.000   Max.   :199.0    Max.   :122.00    Max.   :99.00
                  NA's   :5        NA's   :35        NA's   :227
    insulin            mass            pedigree           age
 Min.   : 14.00   Min.   :18.20    Min.   :0.0780    Min.   :21.00
 1st Qu.: 76.25   1st Qu.:27.50    1st Qu.:0.2437    1st Qu.:24.00
 Median :125.00   Median :32.30    Median :0.3725    Median :29.00
 Mean   :155.55   Mean   :32.46    Mean   :0.4719    Mean   :33.24
 3rd Qu.:190.00   3rd Qu.:36.60    3rd Qu.:0.6262    3rd Qu.:41.00
 Max.   :846.00   Max.   :67.10    Max.   :2.4200    Max.   :81.00
 NA's   :374      NA's   :11
```

```
diabetes
neg:500
pos:268
```

For factors the output is again the number of elements per category.

The summary() function is a very useful tool which can be used on many different kinds of objects. On a dataset, like PimaIndiansDiabetes2, we can rapidly have an idea on the values of the distribution, mainly where the center of the variable is and the presence of possible outliers. For example, looking at variables pregnant and pedigree we see that values between the minimum and $Q3$ are in a small range, with much larger values (their maxima) far from these ranges. Another informative output is the high number of NAs found for variables triceps and insulin, which questions the reliability of these data.

5.1.2 Dispersion Measures

The position measures from the previous section give us an idea about where to find representative values of the sample. This expresses a sense of location, which can be used to interpret, for example, the most likely values. However, characterizing a data sample, this is learning from it, requires of much more measures and calculations to fully describe it. As it can be seen in Fig. 5.2 we have two samples with identical mean, median, and mode values, and yet, the samples are clearly different.

If we want to distinguish between both samples we need a new descriptor. The second level of measures that we can calculate on a sample are the *dispersion measures*, which express the level of data scattering with respect to a central position, this is, how diffuse our values are compared to a value of reference, such as the mean or the median.

One of the most basic dispersion measures is the *range* of the sample, which we have already mentioned. It is simply the distance between the sample maximum and minimum values: $Range(X) = max(X) - min(X)$. This distance merely tells us the extend of maximum separation between observations. The range can be strongly affected by the presence of an outlier, giving us a wrong idea of the data dispersion if most of the observations are located around a central position but the maximum or the minimum is located far from there. We can easily calculate the range using functions `min()` and `max()` or function `range()` as seen in Sect. 2.2.2.

A simple solution for the range's strong dependency on outliers is the *interquartile range* (IQR). If the central positions of a variable are representative we can expect most of the values to be located around them. The IQR calculates the range occupied by the central 50% of observations. We calculate the IQR as the difference between the third and the first quartile: $IQR = Q3 - Q1$. This is a measure easy to interpret: given the IQR, we have a rough idea on how disperse this central mass of values is and, in addition, it is more resistant than the range to outliers.

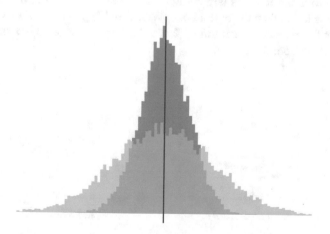

Fig. 5.2 Variables with the same central positions and different dispersion

The IQR can be easily calculated in **R** with function IQR():

```
IQR(mass, na.rm=TRUE)
```

```
[1] 9.1
```

Let us see what we can learn from our dataset

```
sapply(PimaIndiansDiabetes2[, -9], IQR, na.rm=TRUE)
```

```
pregnant  glucose  pressure   triceps
  5.0000  42.0000   16.0000   14.0000
 insulin      mass  pedigree       age
113.7500    9.1000    0.3825   17.0000
```

Variables with higher dispersion will show higher interquartile ranges, but we must remember that this measure depends on the units used to express the variable, and the IQR from two different variables might not be comparable.

Since the IQR uses the central 50% of observations it can be used to define an extreme value. It is commonly considered that extreme values are those smaller than $Q1 - 1.5 \cdot IQR$ or greater than $Q3 + 1.5 \cdot IQR$, those are the values we call *outliers*. These values are atypical, and can be due to an observational mistake, or just abnormal measures of a variable compared with the rest.

Quartiles and the interquartile range can be graphically represented with a *box plot*, also known as *box-and-whiskers* diagrams, as introduced in Sect. 4.1.2. This is an essential tool in any descriptive analysis that summarizes the most important position measures at a glance, namely the median, the quartiles, the maximum and minimum values, and the outliers. The following call produces the diagrams in Fig. 5.3:

```
boxplot(mass, xlab="Body mass index", col="papayawhip")
boxplot(pressure, xlab="Systolic blood pressure", col="pink")
```

The mass variable box plot shows several outlier values at the top of the image, indicating a long tail towards larger values, as we saw in Fig. 5.1. In addition, the top whisker is larger than the bottom one. With variable pressure instead we see a symmetric figure, with the central 50% of values tightly centered around the mean and large tails in both directions.

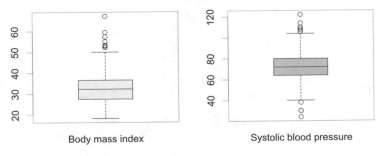

Fig. 5.3 Box plots for variables mass (left) and pressure (right)

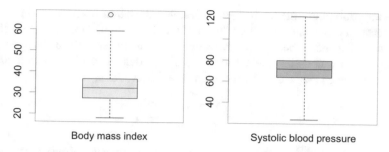

Fig. 5.4 Box plots for `mass` (left) and `pressure` (right) with `range` equal to 3

Note that we can explicitly obtain the outliers by using the `out` method of the function `boxplot()`,

```
boxplot(mass)$out
```

```
[1]  53.2 55.0 67.1 52.3 52.3 52.9 59.4 57.3
```

```
boxplot(pressure)$out
```

```
[1]   30 110 108 122  30 110 108 110  24  38 106 106 106 114
```

The threshold for deciding what should be considered an outlier can be shifted here. If, for example, we want to be more flexible with them and consider that an outlier is any value which is more than 3 times the IQR away from $Q1$ or $Q3$, we can set the argument `range=3` (by default is 1.5). The new box plots in Fig. 5.4 are produced with the code

```
boxplot(mass, range=3, xlab="Body mass index", col="papayawhip")
boxplot(pressure, range=3, xlab="Systolic blood pressure",
        col="pink")
```

and now it turns out that there is only one outlier for `mass` and no outliers for `pressure`.

```
boxplot(mass, range=3)$out
```

```
[1]  67.1
```

```
boxplot(pressure, range=3)$out
```

```
numeric(0)
```

5.1.2.1 The Variance

One of the first problems in modern statistics was the study of errors [6]. If we understand the mean of a variable as its *expected value* (or *expectancy*), as it is defined in probability theory in Sect. 5.2.2, any deviation from that value can be considered an error. The degree of fitting of all observations to the mean must be a measure of dispersion, being smaller when values are closer to the expectancy.

This measure is called the *variance*, and it is defined as the squared sum of distances between the arithmetic mean and the observations:

$$S^2 = \frac{\sum_{i=1}^{n}(x_i - \overline{x})^2}{n}.$$

The quantity $x_i - \overline{x}$ is the error or distance between each observation and the average, and their sum should satisfy the intuitive idea of dispersion. However, the mean is defined as the value that minimizes the distance to each observation and, as a consequence, the sum of all errors is 0. When measuring the dispersion we are not interested in the direction of the possible deviancy, but on its magnitude. For this reason distances must be treated as positive numbers and we square them. Finally, the sum is divided by n, averaging the resulting quantity and giving a precise sense of dispersion. Notice how, given this definition, the variance is a non-negative value, being zero only when all values in the sample are equal.

A natural alternative is to use absolute values instead of squares, as in

$$\frac{1}{n}\sum_{i=1}^{n}|x_i - \overline{x}|$$

(see Exercise 5.6). However, the absolute value function presents problematic mathematical properties.[7] For this reason, this quantity, sometimes referred as the mean absolute error with respect to the mean, is usually dismissed in favor of the variance.

Nevertheless, squaring the distances overestimates the impact of the highest values. This justifies the definition of the *standard deviation*, the square root of the variance, $S = \sqrt{S^2}$. This is one of the most fundamental and widely used measures in statistics, as we will keep seeing on any further section.

Similarly as we have defined the mean absolute error, we can define any dispersion measure with respect to a central position, being the mean, the median, or the mode. The mean is the most common option, so we will dispense with other alternatives, less common in statistics.

For some statistical reasons that will be presented later (see Sect. 5.2) it is sometimes desirable to compute the *unbiased variance* and the *unbiased standard*

---

[7]For example, not being differentiable at $x = 0$.

deviation which are the same sum of squared distances but divided by $n - 1$ instead of n,[8] that is,

$$S_1^2 = \frac{\sum_{i=1}^{n}(x_i - \overline{x})^2}{n-1}, \qquad S_1 = \sqrt{S_1^2}.$$

Note that, for large values of n, the number of elements, both the variance and the unbiased variance, tends to the same value. These two are so important inside statistical inference that the built-in function in **R** to calculate the variance and the standard deviation, `var()` and `sd()`, are defined as the unbiased versions. Still, in many situations, as we will see in the forthcoming sections, we might be interested on having a function for the original variance and standard deviation.[9] As it was mentioned in the introductory Chap. 1, **R** is an exhaustive statistical software and it tends gradually to include many functions, no matter how specific they can be. Usually, the analyst can rely on the existence of an implementation of any algorithm or method since, if it is not in the basic libraries, plenty of developers are constantly adding packages with new tools. Nonetheless, sometimes a particular function is not around because it is really new or maybe for some other mathematical reasons (such as being so similar to an implemented one, like the biased and unbiased variances). In such cases, we can still write the code ourselves.

Recall from Sect. 2.4 how to create our own functions in **R** and then note that $S^2 = \frac{n-1}{n}S_1^2$. We can define these functions as:

```
var.p <- function(x){
         var(x, na.rm=TRUE) * (length(x) - 1) / (length(x))
}
sd.p <- function(x){
         sqrt(var.p(x))
}
```

A toy model can be proposed in order to fully understand the variance and standard deviation. Imagine two classrooms of ten students each, in a grading system that ranges from 0 to 10. The grades in each class have been the following:

```
class1 <- c(5, 5, 5, 5, 5, 5, 5, 5, 5, 5)
class2 <- c(0, 0, 0, 0, 0, 10, 10, 10, 10, 10)
```

In both classes the arithmetic mean is $\overline{x} = 5$. However, in `class1` all data are close to the mean (in fact, all of them equal the mean) but in `class2` all observations

---

[8]This modification in the denominator is the so-called Bessel's correction, which is also attributed to Gauss, see [15].

[9]Usually called population variance and population standard deviation.

are far from it. We can use dispersion measures to see a clear difference between samples. Using our functions we can obtain

```
var.p(class1)
```

```
[1] 0
```

```
var.p(class2)
```

```
[1] 25
```

```
sd.p(class1)
```

```
[1] 0
```

```
sd.p(class2)
```

```
[1] 5
```

Here, it is clearly appreciated how the standard deviation adopts the same scale of the observations, as the difference between each value and the mean is always 5 and thus, this is also the value of the mean.

The example of the two classes is a very particular one, where both populations can be compared in terms of dispersion as they are in the same scale. Going back to the diabetes dataset, we can calculate the variance of variable mass.

```
var.p(mass)
```

```
[1] 47.89302
```

```
sd.p(mass)
```

```
[1] 6.920478
```

The obtained standard deviation is slightly smaller than the IQR, and can be used to compare several variables:

```
sapply(PimaIndiansDiabetes2[, -9], sd.p)
```

```
     pregnant      glucose     pressure       triceps
    3.3673836   30.5157546   12.3740943    10.4701592
      insulin         mass     pedigree           age
  118.6985020    6.9204784    0.3311128    11.7525726
```

The obtained result expresses again similar conclusions to those obtained with the IQR, although slightly smaller. Could we answer which one has a greater dispersion? Here we face the problem of comparing two measures of dispersion that have different units. For example, the body mass index is measured in kilograms by squared meter, and the pressure in millimeters of mercury, and hence it makes no sense to compare them straightforward. In order to do it, we define a relative measure of dispersion, the *coefficient of variation* (*CV* for short), as the standard deviation divided by the absolute value of the mean, that is,

$$CV = \frac{S}{|\overline{x}|}.$$

In **R** we can compute the coefficient of variation by dividing the corresponding values:[10]

```
sd.p(mass) / mean(mass, na.rm=TRUE)
```

```
[1] 0.2132169
```

```
sd.p(pressure) / mean(pressure, na.rm=TRUE)
```

```
[1] 0.1709007
```

Hence, we know that the dispersion in the variable `mass` is greater than in the variable `pressure`. More specifically, `mass` deviates 21.2% from its arithmetic mean on average, while `pressure` does 16.7%. There is something else we can say about the coefficient of variation. As a relative measure for the dispersion in terms of the mean, it also measures the *representativity* of the latter in absolute terms. The coefficient of variation can vary from 0 to the square root of the sample size, $\sqrt{n-1}$ (see [14] for a proof), although big values are unusual and come together with the presence of outliers (see Exercise 5.7). The smaller the coefficient is, the more representative the arithmetic mean is for the whole distribution. In summary, the CV provides a way of comparing the dispersion of two variables in different scales.

5.1.3 Shape Measures

Position and dispersion measures can give us an idea of how values of a variable are structured. Nevertheless, they do not completely determine the distribution of the values. We can have very different distributions with similar position and dispersion measures. These can be different in terms of the *shape*. In Fig. 5.5 we can see a new example where two populations have the same mean and variance, and yet they are clearly different. We need new ways of distinguishing these datasets which, as can be seen, have the same center, marked by the red line, and the frequencies span the same width, i.e., have the same dispersion.

In Sect. 5.1.1 we already mentioned the concept of *symmetry* and its relation to mean and median. This is our first shape statistic, and a fundamental quantity to characterize any sample. The intuitive idea is that a symmetric variable will show the same frequencies (in the same order and the same amount) at both sides of the

---

[10]We could have also used the function `cv()` in the `raster` package.

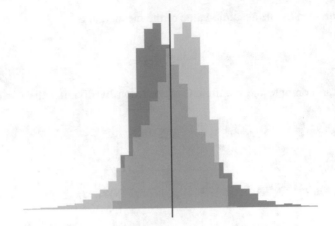

Fig. 5.5 Variables with the same mean and dispersion, but different shape

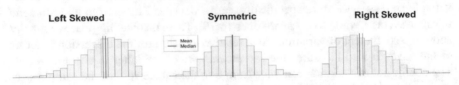

Fig. 5.6 Different skewness types. From left to right: Left skewness, symmetric, and right skewness

mean. In Fig. 5.2 we have two symmetric histograms, where bars at the left side of the red line are a close reflection of those at the right. Red line is the coincident value for the mean and the median, which tells as the relation between these measures and the sample shape: symmetric populations will have coincident mean and median. We must be cautious, tough; it is a common mistake to believe that the opposite is true. We will see examples that this is not the case.

Non-symmetric samples create a particular pattern, a longer tail of frequencies towards one of the possible directions. Therefore, in one dimension, we can distinguish between three different kinds of asymmetry depending on the direction towards the tail extends to: left-tailed, right-tailed, or none (i.e., symmetric). The degree of asymmetry is measured by the *skewness* and we can use it to classify histograms as before. In Fig. 5.6 we have the three possible types of skewness for a unimodal distribution, that is, a distribution with only one mode and hence only one peak. From left to right, we call them left skewed (or tailed), symmetric, and right skewed (or tailed).

As it was the case with central and dispersion measures, skewness can be numerically defined in several ways. Possibly, the most common one is the *Fisher Skewness Coefficient*, defined as

$$g_1 = \frac{\frac{1}{n}\sum_{i=1}^{n}(x_i - \overline{x})^3}{S^3}$$

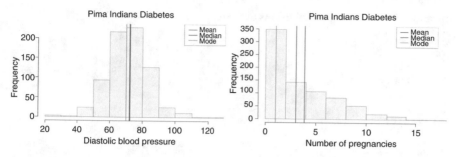

Fig. 5.7 Histogram for variables `pressure` (left) and `pregnant` (right)

which is equal to zero for perfectly symmetrical distributions,[11] smaller than zero for left skewed distributions and greater than zero for right skewed ones. In **R**, we can use the `skewness()` from the `moments`[12] package to compute it.

```
library(moments)
```

Now we will use this new measure to analyze our variables from the sample of the `PimaIndiansDiabetes2` dataset. Recall the box plot for variable `pressure` in Fig. 5.4 right. We can see how the extension of both the box and the whiskers at both sides of the median is approximately the same. This is a sign of a symmetric sample and we should expect $g_1 \simeq 0$. But this dataset contains variables with strong skewness, such as variable `pregnant`.

```
skewness(pressure, na.rm=TRUE)
```

```
[1] 0.133878
```

```
skewness(pregnant, na.rm=TRUE)
```

```
[1] 0.8999119
```

The same conclusion can be easily obtained from Fig. 5.7.

Note that in Fig. 5.6 we have depicted the mean and the median of each variable. For the left skewed distribution we have $\bar{x} \leqslant Me$, $\bar{x} = Me$ for symmetric ones and $\bar{x} \geqslant Me$ for the right skewed. This is an interesting effect of the longer tail over the mean, which is shifted towards the larger or smaller values of the tail. On the other hand, as we saw in Sect. 5.1.1, the median is more resistant to the extreme values of a tail, and this effect is not that strong. For unimodal distributions this is usually the case, although it is not necessary (counterexamples can be found in [22]). For

---

[11]The Normal distribution, that we will see later in Sect. 5.2, is an example of a symmetric distribution.

[12]The name of the package is due to the measure $m_r = \frac{1}{n}\sum_{i=1}^{n}(x_i - \bar{x})^r$, that is, called the *central moment of order r*. Under this perspective, the variance is the second-order central moment, $S^2 = m_2$ and the skewness coefficient is $g_1 = \frac{m_3}{S^3}$.

instance, the following distributions are both right skewed:

```
x <- c(rep(1, 100), rep(2, 140), rep(3, 25), rep(4, 15),
       rep(5, 10))
y <- c(rep(1, 100), rep(2, 140), rep(3, 30), rep(4, 20),
       rep(5, 15))
skewness(x)
```

```
[1] 1.340999
```

```
skewness(y)
```

```
[1] 1.214411
```

but the relative position between their mean and median differs, as shown in Fig. 5.8.

Moreover, a *long* tail can be balanced with a short but *fat* one so that we can have distributions with a zero skewness coefficient but clearly non-symmetrical. For example, the following distribution has a zero skewness coefficient:

```
x <- c(rep(1, 30), rep(2, 40), rep(3, 250), rep(4, 400),
       rep(5, 150), rep(6, 120), rep(7, 10))
skewness(x)
```

```
[1] 0
```

and yet, it is clearly non-symmetric, as shown in Fig. 5.9.

Fig. 5.8 Relative positions of mean and median for two left skewed distributions

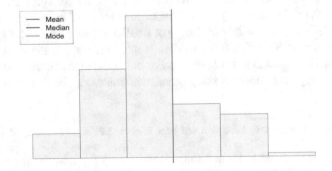

Fig. 5.9 Non symmetric distribution with null skewness coefficient

For distributions with more than one mode, the interpretation is even more difficult. However, for unimodal symmetric distributions we find that the skewness coefficient is always zero, the mode coincides with mean and median, and the median is exactly in the middle of $Q1$ and $Q3$, as we pointed out in Sect. 5.1.1.

The other important shape measure is the *kurtosis*,[13] which measures the relative difference between the tails and the *peakness* of distributions. This can be achieved, in some sense, using the *Pearson Kurtosis coefficient*, which is defined[14] as

$$g_2 = \frac{\frac{1}{n}\sum_{i=1}^{n}(x_i - \overline{x})^4}{S^4}.$$

This number is always positive, and the value is usually compared to 3, which is the kurtosis of the Normal distribution (see Sect. 5.2.2).[15] Similarly as we did with the skewness, we can distinguish three different cases depending on the level of excess. We define a distribution to be *mesokurtic* if the kurtosis is 3, i.e., has the level of kurtosis of a Normal distribution, our reference value. For a kurtosis smaller than 3, we say that the distribution is *platykurtic* and we will call it *leptokurtic* whenever it is greater than 3. We can compute the kurtosis coefficient using the `kurtosis` function in the `moments` package. In the Pima Indian Diabetes dataset, we can compute the kurtosis coefficient for one variable with

```
kurtosis(triceps, na.rm=TRUE)
```

```
[1] 5.897361
```

and then we can affirm that `triceps` is leptokurtic. The rest of variables in the dataset show different values, which can be used to distinguish which one shows a higher kurtosis

```
sapply(PimaIndiansDiabetes2[,-9], FUN=kurtosis, na.rm=TRUE)
```

```
pregnant  glucose pressure   triceps
3.150383 2.716919 3.896780  5.897361
 insulin     mass pedigree       age
9.274775 3.849771 8.550792  3.631177
```

and hence we can compare in Fig. 5.10 two different variables in terms of kurtosis, using their histograms. The interpretation of the kurtosis is not easy and has been susceptible to debate along the years. It has been sometimes considered as a measure of the peakedness, considering than the larger the kurtosis coefficient, the more pointed the distribution is. Nowadays, it seems to be some consensus [24] in the interpretation of the kurtosis as a measure of the size of the tails of the distribution. In this sense, it can be also thought as a measure detecting outliers

---

[13]From Greek *kurtōsis*, "a bulging."

[14]Again, this coefficient can be defined in terms of moments, as $g_2 = \frac{m_4}{S^4}$.

[15]It is actually frequent to talk about excess kurtosis, which is $g_2 - 3$ in order to compare values with zero.

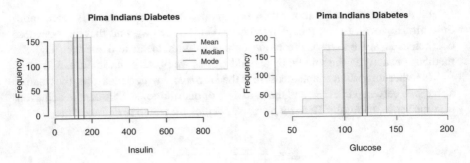

Fig. 5.10 Histogram for `insulin` (left) and `glucose` (right) variables

(see Exercise 5.11). In this vein, we observe how the kurtosis of `insulin` is higher than in `glucose`, given the bigger tail of its distribution (see Fig. 5.10).

In general, shape measures have a difficult and non-unique interpretations. The more measures we have at our disposal, the better the interpretation, and they should be used carefully to obtain a global idea of the distribution at hand. In terms of bounds, the absolute value of the skewness coefficient is always smaller than $n - 1$ where n is the number of the elements of the distribution. For the kurtosis coefficient, it is always smaller than n. Both statements can be found in [13].

5.1.4 Concentration Measures

When dealing with variables which are modelling monetary quantities in economy, subject to be shared by certain people, it is very common to compute measures that inform us about the *concentration* of money in the hands of individuals. This is very useful when we are interested in the division of some goods into several agents and it is usually used as a measure of wealth in an economic context. The most important concentration measure is the *Gini index* introduced by [7] as, in his own words, "the mean difference from all observed quantities." We can define the Gini index for n elements $x_1, x_2, \ldots, x_n$ (which define the amount of wealth going to each individual) as

$$G = \frac{\sum_{i,j}^{n} |x_i - x_j|}{2n^2 \overline{x}}.$$

This index measures the concentration or inequality in the distribution of the frequencies of the different values. On the one hand, if the values were perfectly equally distributed, the numerator of this formula is zero (that is, every person or country has the same amount of the variable at hand) and hence the Gini index. On the other hand, if the concentration is maximal, that is, one individual has the total of the goods (so one value accumulates the 100% of the values of the distribution) and the rest get zero, then the Gini index is close to one.[16] Every value in between indicates more or less inequality or concentration. The values in this interpretation are non-negative (which is the case when they represent an economic good or a sign of wealth in some sense), but the Gini index formula makes sense for negative values and could also be negative itself. For a better interpretation of this index, it is useful to introduce a toy example in an economical context. Let us imagine two Artificial Intelligence start-ups with the following salaries for their employees, in thousands of euros:

```
salary1 <- c(10, 15, 20, 21, 22, 35, 35, 38, 60, 100)
salary2 <- c(21, 22, 23, 25, 32, 38, 42, 46, 46, 50)
```

How can we measure the degree of equality of the distribution of salaries in each company? How can we compare them in these terms? We use the Gini index with the function Gini() in package ineq[17] and obtain the following indices:

```
Gini(salary1)
```

```
[1] 0.3567416
```

```
Gini(salary2)
```

```
[1] 0.1742029
```

That is, the second company has more equally distributed salaries among their employees, although the CEO of the first one probably is not very concerned about it (see Exercise 5.13 for a possible solution). Although the traditional use of Gini index focuses on measuring the inequality or distribution of wealth among the individuals, it can also be used as a measure to detect concentration of values in any context as, for example, decision trees and other machine learning techniques [4]. This way, we can use it to detect more or less concentration in the variables of our dataset, for example, comparing

```
Gini(pregnant, na.rm=TRUE)
```

```
[1] 0.4802395
```

```
Gini(pressure, na.rm=TRUE)
```

```
[1] 0.09467328
```

we see that the variable pregnant is much more concentrated than the variable pressure. This is, in comparison with the distribution of blood pressure, the figures of women pregnancies are much less spread out.

---

[16]Exactly $1 - \frac{1}{n}$, where n is the number of elements of the distribution.

[17]From *inequality*. There are more packages including the computation of the Gini index, such as lawstat or reldist. The one proposed by the authors uses the formula $G = \frac{2\sum_{i=1}^{n} i x_i}{n \sum_{i=1}^{n} x_i} - \frac{n+1}{n}$, where $x_1 \leqslant x_2 \leqslant \ldots \leqslant x_n$, which is equivalent to the one stated in the text.

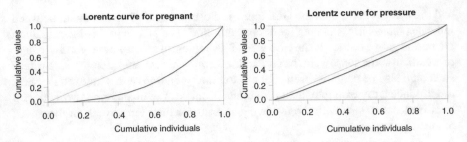

Fig. 5.11 Lorentz curves for pregnant (left) and pressure (right) variables

There is a graphical alternative to represent the concentration of a set of numbers, the *Lorentz curve*. It represents the cumulative distribution of the variable for the cumulative set of individuals, accounting for how much of the total variable sum (to be shared) is distributed for each percentage of the population over which the variable is measured. We can plot it with the package ineq using function Lc(). The function itself computes and stores the cumulative values of the variable and the individuals to plot the curve. We draw the Lorentz curve of the previous two variables, to see the differences

```
plot(Lc(pregnant), col="red", lwd=2,
    main="Lorentz curve for pregnant",
    xlab="Cumulative individuals", ylab="Cumulative values")
plot(Lc(pressure, na.rm=TRUE), col="blue", lwd=2,
    main="Lorentz curve for pressure",
    xlab="Cumulative individuals", ylab="Cumulative values")
```

and we obtain the plots in Fig. 5.11. As we already know, the variable pregnant is much more concentrated than pressure. The interpretation of the curve is that each point (p, q) depicts the percentage q of values accumulated by the percentage p of those individuals getting the lowest. To plot this curve for a distribution with (ordered) values $x_1 \leqslant x_2 \leqslant \ldots, \leqslant x_n$, the corresponding points are $p_i = \frac{i}{n}$ and $q_i = \frac{\sum_{j=1}^{i} x_j}{\sum_{j=1}^{n} x_j}$ for $i = 1, 2, \ldots, n$. We use these points together with $(0, 0)$ to plot the curve as a piecewise linear curve joining these points. Note that the diagonal $p_i = q_i$ is interpreted as the perfect equality in the distribution, all individuals share the same amounts. That means, the closer the Lorentz curve is to the diagonal line, the more equally distributed the values among the individuals are. Also, note that the curve is always below the diagonal since we order values from lowest, and hence accumulation of values cannot be greater than the one of individuals.

The Lorentz curve contains more information about the concentration of the distribution since it does not only contain the global concentration but also how this concentration is scattered along the list of individuals. Hence, two completely different Lorentz curves could correspond to the same Gini index (see Exercise 5.14).

It turns out that the Gini index can be derived (and it is in fact the way some textbooks define it) from the Lorentz curve in terms of areas: the Gini index is the area between the diagonal and the curve divided by the total area under the diagonal.

This is shown to be equivalent to the definition above. It can also be computed as

$$G = \frac{\sum_{i=1}^{n-1} p_i - q_i}{\sum_{i=1}^{n-1} p_i}$$

There are more indices to measure concentration. Again, in the **R** package `ineq` we can find the *Theil index*, a measure of concentration based in the *entropy* of distribution of numbers which, in turn, measures the degree of chaos and spreadness of the numbers. The formula for the Theil index of n numbers $x_1, x_2, \ldots, x_n$ is given by

$$T = \frac{\sum_{i=1}^{n} x_i \log \frac{x_i}{\bar{x}}}{n\bar{x}}$$

and the coefficient varies between 0 and $\log n$. For example, this index for the variables `pregnant` and `pressure` takes the values

```
library(ineq)
Theil(pregnant)
```

```
[1] 0.2476443
```

```
Theil(pressure)
```

```
[1] 0.01478803
```

and these results offer a similar interpretation to Gini index, but in a different scale.

5.1.5 Exercises

Exercise 5.1 Compare the arithmetic mean, the median, and the mode of the `triceps` variable in the Pima Indian Diabetes dataset. Plot a histogram to show the three position measures. Conclude, based on Sect. 5.1.3, what the skewness of the distribution should be.

Exercise 5.2 Compute the geometric and harmonic means of `triceps` by using the mathematical formula described in Sect. 5.1.1. Compare the values with the results of applying the built-in functions `geometric.mean` and `harmonic.mean` from package `psych`.

Exercise 5.3 Construct a function in **R** (as in Sect. 2.4) that computes the mode of a set of numbers in the unimodal case. If there is no mode, the function must show a warning message declaring it.

Exercise 5.4 For the `pressure` variable in `PimaIndiansDiabetes2`, compute the highest value which is smaller than the 35% of the data. Also, compute the minimum blood pressure of the 10% women with the greatest blood pressure.

Exercise 5.5 Plot a histogram for the `glucose` variable with a bar comprehending between quartile values. Show each bar in a different color.

Exercise 5.6 Consider the formula $\frac{\sum_{i=1}^{n}(x_i-\overline{x})}{n}$, as a candidate to measure dispersion and test it in **R** over some of the examples in this book. Once you realize the common result to all of them, prove it analytically and explain why this is not a good measure of dispersion. Think of alternatives, apart from, of course, variance and standard deviation, to measure how representative the arithmetic mean of a distribution can be, implement them in **R** and discuss the results. Hint: use the absolute value and even powers of the differences.

Exercise 5.7 Find an example of a variable with coefficient of variation greater than one.

Exercise 5.8 Create a univariate dataset corresponding to measures of a variable, and add some outliers to the distribution. Show that the median absolute deviation is less sensitive to them than the standard deviation (as it was explained to compare the arithmetic mean and the median in Sect. 5.1.1).

Exercise 5.9 Compute the maximum and minimum coefficients of skewness for the variables in the Pima Indian Diabetes dataset and decide which one is more skewed and which one is more symmetric.

Exercise 5.10 Define variables and datasets which qualitatively reproduce the histograms in Fig. 5.6. Check their skewness coefficients and the relative positions of mean, median, and mode to support the matching of the variables with each picture.

Exercise 5.11 Define a variable with numerical sample of values and compute its kurtosis coefficient. Introduce new elements, such as numbers similar to the arithmetic mean, median, or mode, numbers that enlarge the distribution tails or outliers, and determine the change that takes place in the coefficient. Interpret the changes in the distribution, through comparing bar plots or histograms.

Exercise 5.12 The four authors of a book want to split the selling earnings of the book in a way that they make more money depending on the working time that they spent in the project. They are ordered increasingly in time invested, with four different fractions of time. Calculate the investing times such that the splitting of the 50,000 euros that they earned yields a Gini index equal to 0.15.

Exercise 5.13 Try to change as few data salaries as possible in the first company of the toy example of Gini index (Sect. 5.1.4), in order to obtain a more equal distribution of salaries than the second company keeping the same total amount of salaries.

Exercise 5.14 Create an example of concentration calculation for two different splittings with, approximately, the same Gini index but different Lorentz curves.

Exercise 5.15 According to some organizations, the world is so uneven that the 1% of the population own the 82% of the wealth. Compute the Gini index for this situation.

Exercise 5.16 Compute the Gini and Theil indices of the numerical variables in the Pima Indian Diabetes dataset and store them into two vectors. Compare whether one index is below or above the other, for each variable.

5.2 Statistical Inference

Previously in this chapter we have performed analysis over some of the variables in the Pima Indian Diabetes dataset. This analysis allows us to understand how these measures are distributed along the possible values. In that dataset, we have data coming from 9 variables measured in 768 women. But there are many more women in the Pima tribe which could have been taken into account. In statistics, obtaining large samples from a population is always desirable, but not always possible.

Why would one be interested in analyzing only this subset of the tribe? What is it useful for? In this particular case, the reason is to understand, relate, and extract conclusions about the values of these physical measures, in order to prevent some diseases or health behaviors using these variables. More precisely, the ultimate reason to study this dataset is to be able to predict the diabetes disease using the rest of the variables.

In order to do that we would like, ideally, to have information about the whole *population*, in this case the whole Pima tribe. However, we do not usually have time, space, tools, or license to measure all variables in the whole population. Most of times we can only obtain information about some individuals, called a *sample*, to carry out our studies. With this partial information we would like to derive conclusions about the whole set of individuals and here is where *statistical inference* takes part. Ideally, with the data from the 768 women we would like to understand the distribution of the values of all variables in the whole set of women. We use the sample of 768 observations to *infer* conclusions about the whole population.

5.2.1 Random Variables

5.2.1.1 Probability Theory

In order to model measurements of a variable in a whole population we shall use probability theory. We only introduce the idea and basic definitions of probability without deepening into its study in order to arrive to the concept of random variable. We all share the idea of probability as a number indicating our degree of uncertainty or belief about some fact or event. Mathematically modeling probability, though, is a slightly more involved story. There are various ways of interpreting what probability is from which we mention the main three.

The *classic* notion of probability, mainly due to Laplace [20], focuses only on the simple case when all possible events of a random experiment have the same chances of happening. Then, the probability of some event is defined just with the rule of dividing the number of possible successful outcomes over the number of total possible outcomes. For example, if we roll a die, it seems clear that we have a probability of $\frac{1}{2}$ of obtaining an even number, because there are three favorable cases, 2, 4, and 6, out of a total of 6 possible cases.

Another approach is the *frequentist* conception of probability [9, 21, 23], in which we would roll the die many times and we will observe the number of times we obtain an even number out of the total number of rollings. The probability will be, thus, the limit of this frequency when the number of experiments tends to infinity.[18]

Finally, we should mention the *subjective* or *Bayesian* interpretation of the probability [3, 12, 17, 18], where users assign probabilities *a priori* based on their beliefs and available information, and therefore, two different users may conclude different probabilities, being both correct from their respective points of view. Bayesian probability is developed from this approach and it is a powerful tool in statistics nowadays.

Although there are different interpretations of probability,[19] the mathematical modelization of this concept is unique and corresponds to an axiomatization due to the Russian mathematician Andrey Kolmogorov in 1933 (see [16] for an English translation). We begin with a set of possible outcomes Ω for a random phenomenon. For instance, imagine that we draw a card from a standard 52-card deck that will be denoted by Ω. Now, consider every possible subset of outcomes that could take place, such as drawing a *queen* or that the card is a *spade*. This set of outcomes is called the *sample space*[20] S and we call its elements *events*.[21] We can consider unions $A \cup B$, such as the card being a *heart* or a *diamond* (hence, any red card) or intersections of events $A \cap B$, such as being a *queen* and a *spade* at the same time (hence, the queen of spades). We will say that two events A and B are *incompatible* if whenever one occurs, the other is impossible, then $A \cap B = \emptyset$. An example for this is being a *club* and being a *diamond*, there is no card that can match both at the same time. Two events are called *independent* when the probability of one taking place does not affect the probability of the other. For example, if we take two cards of the deck at the same time, the probability of being a diamond or a club is different for the first and the second cards since the result of the first affects the second. This

---

[18]Mathematically this is expressed with a limit: $P(A) = \lim_{n \to \infty} \frac{n_A}{n}$, the probability of occurring an event A is the limit of the quotient between the number of trials with an A outcome, over the total number of trials, when this number goes to infinity.

[19]For an extensive explanation the different schools of probability we recommend the prologue of [3].

[20]The technical definition of this and other incoming ones is out of the scope of this book. In this case, we would need to talk about Borel σ-algebras and measure theory. See [1] for details.

[21]In the example, S would be the power set of Ω, $\mathcal{P}(\Omega)$, the set of all possible subsets of cards.

is not true if we select two cards sequentially with *replacement*, that is, returning the first card to the deck before extracting the second, and they are independent events in this case. Also, we can consider the complementary of an event $A^C = S \backslash A$ as the set of outcomes that are not in the event A.

Then, we define *probability* P as a function assigning to every element A of the sample space S, a real number $P : A \in S \mapsto P(A)$ that satisfies the three *Kolmogorov axioms*, namely:

- For every event A, $P(A) \geqslant 0$.
- The probability for the whole set is $P(\Omega) = 1$.
- The probability of the union of two events with empty intersection $A \cap B = \emptyset$ is the sum of their probabilities $P(A \cup B) = P(A) + P(B)$.[22]

All these elements form a *probability space* (Ω, S, P). From these axioms, we can derive the main properties of probability:

- The probability of any event is between 0 and 1.
- The probability of an intersection of independent events A and B is the product of probabilities $P(A \cap B) = P(A) \cdot P(B)$.
- The probability of the complementary event is $P(A^C) = 1 - P(A)$.

These properties can lead to many others, for example, to compute a surprising probability in Exercise 5.17.

5.2.1.2 Discrete and Continuous Random Variables

After presenting the foundations of probability, we can introduce a new central concept: the *random variable*. We say a function $X : S \longrightarrow \mathbb{R}$ is a random variable if it assigns real numbers to each possible event from S. This way, every event is contained in our probability space (Ω, S, P) and we evaluate its probability.[23] For example, we can roll two dice and study the resulting sum. If the outcomes are 3 and 4, we assign the sum of these numbers with a random variable X and, afterwards, calculate the probability of obtaining a 7 with the probability function P. Random variables are either *discrete* or *continuous*. For example, a random variable counting the number of car accidents in a city for some period of time is a discrete one, since we can only have a natural number of accidents but fractions cannot be considered, as there is no such thing as a "quarter of accident." On the other hand, a random variable measuring the temperature of an airplane engine is a continuous random variable because it can take any value in a certain interval. Technically, we say that discrete variables can only take values from a set of isolated quantities, with a positive distance between values; and continuous variables can take any value in

---

[22]The statement of this third axiom involves (possibly infinite) countable unions of disjoint events. Here we simplify the situation for reader's clarity.

[23]For technicalities and completeness, refer to [1].

an interval. Random variables are the mathematical concept assigning quantities to the different events in a population, i.e., measuring the reality of a *population*. Later, we will assign probabilities to those quantities.

In the Pima Indian Diabetes dataset, we have a *sample* of 768 values from the population which is the set of all possible women's mass in the tribe. Several variables are studied and each one can be regarded as a sample obtained from a random variable. Both mass and pregnant are considered as numeric variables for **R**

```
class(mass)
class(pregnant)
```

```
[1] "numeric"
[1] "numeric"
```

but mass is continuous and pregnant is discrete (2.5 pregnancies are not possible!) and the possible values can be inspected with

```
sort(unique(pregnant))
```

```
[1]  0  1  2  3  4  5  6  7  8  9 10 11 12 13 14 15 17
```

where unique() extracts all different values and, then, they are sorted out with sort().

Once the random variable X is defined we can use real functions to obtain probabilities. The probability function P assigns then probabilities between 0 and 1 to the real values.

$$X : S \longrightarrow \mathbb{R} \qquad\qquad P : \mathbb{R} \longrightarrow [0, 1]$$
$$A \longmapsto X(A) = x \qquad x \longmapsto P(x)$$

For discrete random variables, the function P receives the name of *probability mass function*. However, the same function cannot be built for continuous variables. If X is a continuous random variable, then S is an interval or a union of intervals, containing infinitely many different values. In order to satisfy Kolmogorov's second axiom, we need a non-negative real-valued function $P(X(A) = x) = f(x)$ such that

$$P(S) = P(-\infty < x < \infty) = \int_{-\infty}^{\infty} f(x) = 1 \tag{5.1}$$

and given a continuous random variable X, the probability of giving values in an interval $[a, b]$ is computed as the area below the curve of $f(x)$

$$P(a \leqslant X \leqslant b) = \int_{a}^{b} f(x)dx.$$

With this definition, the probability of obtaining exactly the value x could be defined as the integral over the interval $[a, b]$ where both a and b are equal to x, this is, an interval containing only x. However, the integral over a single value x is always 0, i.e., the probability of obtaining exactly that value from a set of infinitely many possibilities. For example, in a random variable measuring heights of European people, it does not make sense to ask about the probability of being exactly 177.2567 cm tall. Instead, we will consider which is the probability of finding an individual between 175 and 180 cm, which is the probability of an interval. For this reason, the function f is called a *probability density function*, since it provides densities but no pointwise probabilities.

For both, discrete and continuous random variables X, we can define the *cumulative distribution function*[24] as the function assigning to each possible value x the probability of the random variable X taking a value less or equal than x, that is,

$$F(x) = P(X \leqslant x) \,.$$

The way we compute this function from the mass or density functions is, for discrete and continuous cases,

$$F(x) = \sum_{y \leqslant x} f(y) \ \text{ and } \ F(x) = \int_{-\infty}^{x} f(y) dy \,,$$

respectively.

We now graphically represent these functions in an example. Imagine we have a loaded die, with some outcomes more likely than others. We define a discrete random variable assigning to each face its corresponding number, this way, the face with one dot on it is assigned to number 1, the one with two dots is assigned to number 2, and so on. The probability mass function can be seen in Fig. 5.12.

```
loaded <- data.frame(values=c(1, 2, 3, 4, 5, 6),
                     probs=c(1 / 10, 1 / 10, 3 / 10, 2 / 10,
                            2 / 10, 1 / 10))
library(ggplot2)
ggplot(data=loaded, aes(x=values, y=probs)) +
  geom_bar(stat="identity", fill="aquamarine3") +
  ylim(c(0, 0.35))
```

Recall that the sum of the given probabilities is 1, satisfying one of the Kolmogorov axioms. For continuous random variables, we need a probability density function, defined on an interval and not by a finite sequence of values. These

---

[24]Usually called just the *distribution function*.

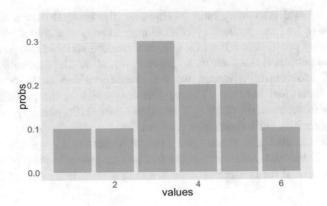

Fig. 5.12 Mass function for the loaded die

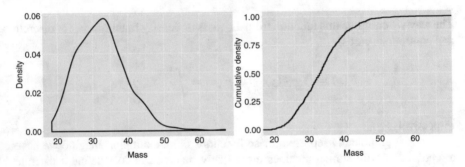

Fig. 5.13 Probability (left) and cumulative (right) density functions for mass

are mathematical functions satisfying equation (5.1), such as the piecewise defined function

$$f(x) = \begin{cases} 0 & x \leq 0 \\ 3x^2 & 0 < x < 1 \\ 0 & 1 \leq x \end{cases}$$

whose integral between $-\infty$ and ∞ is 1. However, we can also estimate probability density functions from real datasets. For example, we can think of variable mass from the Pima Indian dataset as a continuous random variable (recall that it is actually a sample of this variable). We can plot its *empirical* probability density function and cumulative density function by smoothing and rescaling its histogram, with ggplot2 functions geom_density() and stat_ecdf() as follows (see Fig. 5.13):

```
ggplot(data=PimaIndiansDiabetes2, aes(x=mass)) +
  geom_density() +
  xlab("Mass") +
  ylab("Density")
```

```
ggplot(data=PimaIndiansDiabetes2, aes(x=mass)) +
  stat_ecdf() +
  xlab("Mass") +
  ylab("Cumulative density")
```

Although the technicalities are out of the scope of this book, we present a fast way of obtaining empirical estimations of functions $f(x)$ and $F(x)$ for samples of continuous random variables. Function `density()` from package `stats` estimates the probability density function similarly as `ggplot2` does, in order to create Fig. 5.13 left. With the call

```
pdf.density <- density(mass, na.rm=TRUE)
```

we obtain a `density` class object which we will use as a data frame. It contains, among others, vectors `pdf.density$x` and `pdf.density$y`, containing values x and $f(x)$, respectively. For example, the element 150 of these vectors contains the estimated probability density of the value $x = 30.43$, which is around $f(x) = 0.054$:

```
pdf.density$x[150]
```

```
30.42878
```

```
pdf.density$y[150]
```

```
0.05387463
```

Similarly, and from the same package, we can use function `ecdf()` to obtain the empirical cumulative distribution function, which returns, for a given value x, the probability of finding smaller values, $F(x)$. For example, let us compute the percentage of women with less `mass` than 35 is a 67.8%:

```
ecdf.cumul <- ecdf(mass)
ecdf.cumul(35)
```

```
[1] 0.677675
```

5.2.1.3 Measures of Random Variables

For every random variable we can define measures that allow us to understand their distribution. These are the theoretical counterpart of the measures seen in Sect. 5.1 and can be used as well to describe and characterize a random variable. It is important to understand that we assign a random variable to a random population, being the former a mathematization of the later. Therefore, every property of the random variable will have its impact on the population. Through population sampling and descriptive statistics we try to discover the characteristics of the parent random variable. For this reason, it is important to understand the probabilistic equivalent of the measures explained in the previous section.

The most important measure for a random variable X is the *expected value* (also called the *expectancy* or just the *mean*) $E(X)$, which can be estimated from a sample by using the arithmetic mean. It is defined as the sum of the values, weighted by the probability of happening, in the discrete case,

$$\mu = E(X) = \sum_{x \in S} x P(X = x) ,$$

or, for a continuous variable,

$$\mu = E(X) = \int_{-\infty}^{\infty} x f(x) dx .$$

This measure indicates the "center" of the distribution and it is the first *central moment*.[25] As in Sect. 5.1.1, other measures are available for the notion of centrality, such as the median $Me(X)$, defined as the value that accumulates half of the probability of the variable, that is,

$$F(Me(X)) = \int_{-\infty}^{Me} f(x) dx = \frac{1}{2} ,$$

or the mode $Mo(X)$ which is the value (or values) maximizing the mass or density function $f(x)$. If we want to measure the *dispersion* of the possible values of the random variable we shall use the *variance* $V(X)$, defined as $E[(X - \mu)^2]$ and computed as

$$\sigma^2 = V(X) = \sum_{x \in S} (x - \mu)^2 P(X = x) ,$$

and

$$\sigma^2 = V(X) = \int_{-\infty}^{\infty} (x - \mu)^2 f(x) dx ,$$

for X being, respectively, discrete or continuous. Usually, we use the *standard deviation* instead, denoted by $\sigma = \sqrt{\sigma^2}$, because it has the same units than the variable X at hand. These measures obviously correspond to the variance and standard deviation defined in Sect. 5.1.2.

In what follows we will sometimes be interested in transforming a random variable X to a *typified* or *standardized* one, usually written X^*. These are versions

---

[25] Similarly to Sect. 5.1.3, in the continuous case the rth central moment around a point a is given by $\mu_r^a = \int_{-\infty}^{\infty} (x - a)^r f(x) dx$, meaning the average of the distances from the values x to the point a, raised to r. This way the expected value is the first moment about the origin, $E(X) = \mu = \mu_1^0$, and the variance is the second moment around the mean, $V(X) = \mu_2^{E(X)}$.

of an original variable transformed to have mean $\mu = 0$ and standard deviation $\sigma = 1$. This can be achieved by subtracting the mean and dividing the result by the standard deviation,

$$X^* = \frac{X - \mu}{\sigma} .$$

Standardized variables are of great help in statistics. When two variables have the same mean and variance we can compare them to detect differences in higher order moments, such as skewness or kurtosis.

5.2.1.4 Sampling

In statistical inference we assume there is a random variable, associated with the population, from which we want to obtain information. From this population we can obtain *samples*, a finite and hopefully representative subset of the population whose values we can measure and collect. Given a sample, we try to learn as much as possible about the population. There are many ways of obtaining samples from a parent population, but for the sake of simplicity we will mention only the most common one, *simple random sampling (srs)*. Given a population X we select a sample of n elements in which every element has the same probability of being selected, i.e., every subset of n elements of the population has the same probability of becoming our sample. In order to guarantee this independence between individuals we need to sample elements with replacement: every time an element is selected as a member of our sample, we must consider it as eligible again when the next element is selected. This makes samples to contain repeated elements, especially if the sample size is comparable to the population size. In other situations, for example, election polls, it is not correct to repeat observations, since we would be asking twice the same individual.

In **R**, we can obtain samples with the function `sample()`. Given a vector of n elements we can obtain a sample of size m:

```
set.seed(5)
x <- 1 : 50
sample(x, size=10)
```

```
[1] 11 34 45 14  5 32 24 35 41 46
```

The output is 10 elements from x without replacement.[26] For a simple random sampling we must include the argument `replace=TRUE`. In the next example we sample the result of tossing a coin 25 times with two possibilities, heads (H) and

---

[26]Function `set.seed()` guarantees that any **R** user starting his/her code with the same value, `set.seed(n)`, will generate the same random numbers, and therefore identical results.

tails (T). That is, we are simulating a random experiment with 25 trials.

```
set.seed(5)
sample(c("H", "T"), 25, replace=TRUE)
```

```
 [1] "H" "T" "T" "H" "H" "T" "T" "T" "T" "H" "H" "H" "H"
[14] "T" "H" "H" "H" "T" "T" "T" "T" "T" "H" "H" "H"
```

In these examples, we are considering that each possible outcome has the same probability. We can modify these probabilities by using the `prob` option.

In the following example we simulate 10 trials of a tricked die, with different probabilities for each number: 10% chances for results from 1 to 5 and 50% chances of getting a 6.

```
set.seed(5)
sample(1 : 6, size=10, prob=c(0.1, 0.1, 0.1, 0.1, 0.1, 0.5),
       replace=TRUE)
```

```
[1] 6 4 1 6 6 5 3 2 1 6
```

Next section will show how to sample from probability distributions which are widely known and used in practice.

5.2.2   *Probability Distributions*

Section 5.1 dealt with describing properties of distributions in order to ultimately extract information about the population being studied. It turns out that there are some standard distributions in which many of the populations can be fitted. When this is the case, we can focus on studying the distributional properties to acquire a deeper understanding of them. Then, we check whether a population distribution is similar enough to any of these distributions and, thus, assume that our population satisfies the properties of the chosen theoretical distribution. Another way of finding patterns is, even when populations cannot be associated with theoretical distributions, check whether two populations could have been generated by the same underlying distribution or not.

Most of the standard probability distributions arise in random phenomena which are well determined and studied. These are *models* of probability distributions that we use to represent these phenomena. For example, tossing a coin is a very common toy example, since many experiments or situations have two possible outcomes with certain probabilities each, and thus they can be studied in the same way as coin tossing. These distributions are called *parametric* distribution models, since they are completely determined by a set of parameters, for example, the probability of one of the two outcomes in the previous example, or the mean and variance in the case of the normal distribution. The way to proceed, then, is to define probability distributions and study them in general and, after that, use their structure to model our variables.

Table 5.1 R functions for probability distributions from package stats	Discrete		Continuous	
	Distribution	Name	Distribution	Name
	Binomial	binom	Normal	norm
	Negative binomial	nbinom	Student's t	t
	Geometric	geom	Chi squared	chisq
	Poisson	pois	Snedecor's F	F

The use of probability distribution functions in **R** is provided by the package stats which is installed by default. Table 5.1 shows the function calling name for each distribution.

There are more probability distributions in this package, which can be listed with

?Distributions

For each probability distribution there are four functions in **R**. Each of them is called by adding the following prefix to the corresponding distribution's name:

- d—for the mass or density function.
- p—for the (cumulative) distribution function.
- q—for quantiles, that is, to compute the corresponding value for the cumulative distribution function given a probability.
- r—to generate random samples with the given distribution.

Next, let us go over each distribution and explain their meaning.

5.2.2.1 Discrete Probability Distributions

The *Bernoulli* distribution is a discrete random variable with two possible values, 1 and 0, with probabilities p and $q = 1 - p$, respectively. The mass function can be stated as $f(x) = p^x(1 - p)^{1-x}$ for $x = 0, 1$, where the expected value is $E(X) = p$ and the variance is $p(1 - p)$. This is, for example, the distribution of a random phenomenon corresponding to an experiment with two possible outcomes such as tossing a coin.

When we toss more than one coin (or a coin many times) we have the sum of several Bernoulli distributions, each one of these experiments being called a *Bernoulli trial*. For a sum of n independent[27] events, each one with probabilities p (for $X = 1$) and $1-p$ (for $X = 0$), we have the *Binomial* distribution $B(n, p)$, which counts the number of times that $X = 1$ takes place in n trials. Hence $B(1, p)$, a single experiment, recovers the Bernoulli distribution. The mass function of $B(n, p)$ is

$$f(x) = \binom{n}{x} p^x (1 - p)^{n-x},$$

---

[27] Independent meaning that the outcome of one of them do not conditionate the outcome of the following trial.

where $x = 0, 1, \ldots n$. From this, it can be calculated the expected value $E(X) = np$ and the variance $V(X) = np(1 - p)$. For example, suppose that we have 5 women with probability 0.2 of presenting diabetes. The random variable *number of women presenting diabetes* follows a binomial distribution $X \sim B(5, 0.2)$. To compute the probability of having exactly two women with diabetes we use dbinom(), the mass function of the binomial distribution:

```
dbinom(2, size=5, prob=0.2)
```

```
[1] 0.2048
```

If we want the cumulative probability of the distribution for this value, that is, the probability of 2 or less women (i.e., none, one, or two women) presenting the disease, $P(X \leqslant 2)$, this is

```
pbinom(2, 5, 0.2, lower.tail=TRUE)
```

```
[1] 0.94208
```

We can reverse the tail in order to compute the other side of the distribution, probabilities greater or equal than a given number, as $P(X \geqslant 2)$, with

```
pbinom(2, 5, 0.2, lower.tail=FALSE)
```

```
[1] 0.05792
```

We can also calculate the value that accounts for a certain proportion of the cumulative probability by using the quantile function. For example, we can compute x such that $P(X \leqslant x) = 0.95$ (or equivalently $P(X \geqslant x) = 0.05$), that is, the number of women we need to account for, to have a probability of being diabetic equal to 0.95.

```
qbinom(0.95, 5, 0.2)
```

```
[1] 3
```

We can randomly generate elements from a distribution. For example, to draw 10 elements from the binomial $B(5, 0.2)$, we type

```
set.seed(5)
rbinom(10, 5, 0.2)
```

```
[1] 0 1 2 0 0 1 1 2 3 0
```

Note that values range from 0 to 5, and the outcome usually is closer to 0 than to 5, because of the small probability p of getting $X = 1$ in each trial. The following code produces the plots in Fig. 5.14, where we see that the probability of obtaining zero is much higher than the probability of obtaining 5.

```
values <- 0 : 5
plot(values, dbinom(values, 5, 0.2), type="b", xlab="Values",
     ylab="Probabilities", main="Mass")
plot(values, pbinom(values, 5, 0.2), type="b", xlab="Values",
     ylab="Probabilities", main="Cumulative")
```

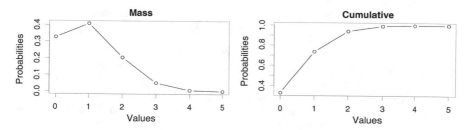

Fig. 5.14 Mass (left) and distribution (right) functions of the Binomial distribution $B(5, 0.2)$

If we count the number of Bernoulli trials with probability p that we need to draw before obtaining the first r events with $X = 1$, we have a *Negative Binomial* distribution $NB(r, p)$, with a mass function

$$f(x) = \binom{x-1}{x-r} p^r (1-p)^{(x-r)} ,$$

yielding mean $E(X) = \frac{r(1-p)}{p}$ and variance $V(X) = \frac{r(1-p)}{p^2}$.

The *Geometric* distribution $G(p)$ is a special case of the negative binomial, for $r = 1$. Therefore, we count, in a sequence of Bernoulli trials with probability p, the number of times needed to obtain the first $X = 1$. It can be used to model the probability of having a specific number of failures until the first success in a two-outcome experiment. Its mass function is

$$f(x) = p(1-p)^{x-1} ,$$

with mean $E(X) = \frac{1-p}{p}$ and variance $V(X) = \frac{1-p}{p^2}$. For example, we can compute the probability of rolling a die 5 times until the first 6 appears. The parameter p of each Bernoulli trial is $p = \frac{1}{6}$, then the probability of 5 failures before a success is given by $f(5) = \frac{1}{6}(\frac{5}{6})^5 = 6.7\%$.

The *Poisson* distribution with parameter λ, written $Po(\lambda)$, is the discrete random variable counting the number of (rare) events happening in a period (of time, space, ...) when these events occur within a known constant rate (and hence independent of the occurrence of the last event). Its mass function is

$$f(x) = e^{-\lambda} \frac{\lambda^x}{x!} ,$$

for $x = 0, 1, \ldots$. The expected value and the variance of a Poisson distribution are both equal to λ.[28] For example, if we know that the number of accidents in an airport is, on average, 5 per year, the probability distribution to model the number

---

[28]The Poisson distribution happens to be the only distribution that satisfies this property.

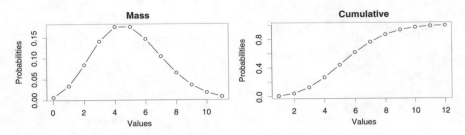

Fig. 5.15 Mass (left) and distribution (right) functions of Poisson distribution $Po(5)$

of accidents in this airport can be a Poisson of parameter $\lambda = 5$, this is $Po(5)$. We plot the mass and cumulative distributions (for just $x = 0, 1, \ldots, 10$)

```
values <- 0 : 11
plot(values, dpois(values, 5), type="b", xlab = "Values",
    ylab="Probabilities", main="Mass")
plot(ppois(values, 5), type="b", xlab = "Values",
    ylab="Probabilities", main="Cumulative")
```

to produce the plots in Fig. 5.15.

The Poisson distribution is a limit case of the binomial distribution, where $n \to \infty$ and $p \to 0$. When the number of trials is large enough but the probability of success is extremely low, we may have a Poisson distribution. Think about accidents over a large population: every day millions of people take their car but the probability of having an accident is, fortunately, very low. However, the number of cars on the streets is so large that it is not strange to have a few accidents every day.

5.2.2.2 Continuous Probability Distributions

The most basic continuous distribution is the one assigning the same probability density to every point of an interval. Namely, consider a real interval $[a, b]$ and a constant non-negative function $f(x) = k \geqslant 0$ for every x in $[a, b]$. If we set the condition that it has to verify in order to be a density function

$$\int_a^b k \; dx = 1 \, ,$$

it turns out that $k = \frac{1}{b-a}$. Hence, the density function is $f(x) = \frac{1}{b-a}$ for $x \in [a, b]$ and this is called the *uniform* distribution. Figure 5.16 shows the graph of this function. It can be easily computed that its expected value is the point in the middle of the interval $\frac{a+b}{2}$, and its variance is equal to $\frac{(b-a)^2}{12}$.

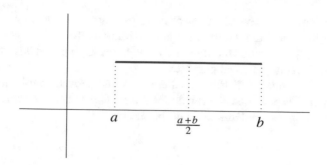

Fig. 5.16 Uniform density function and its expected value

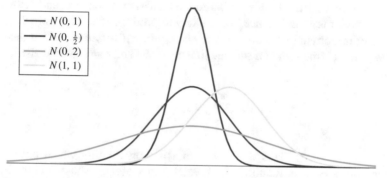

Fig. 5.17 Normal distribution for several values of mean and variance

We shall use this distribution of probability whenever we just know the extreme values that it might take and we do not have more information about it. Then, we assign the same probability to any subinterval within $[a, b]$ with the same length.

The most important continuous distribution by far is the *Normal* or *Gaussian* distribution. It is the distribution arising in many random phenomena in natural sciences. It is also very important because of the Central Limit Theorem that we will discuss later, and which asserts that the normal distribution appears as the limit of other probability distributions. Hence, with large samples we always can approximate any distribution by a normal one. The density function of a normal distribution $N(\mu, \sigma)$[29] with expectancy μ and variance σ^2 is

$$f(x) = \frac{1}{\sqrt{2\pi}\sigma} e^{-\frac{(x-\mu)^2}{2\sigma^2}}.$$

This function has a recognizable symmetric bell shape. Depending on the parameters μ and σ this bell shape will vary. Figure 5.17 shows the density function for the

---

[29]There are two ways to denote the normal distribution: $N(\mu, \sigma^2)$ with the variance, and $N(\mu, \sigma)$ with the standard deviation. We choose the later since **R** notation follows it.

normal distribution for several values of these parameters. Given any $N(\mu, \sigma)$, we can typify it by subtracting the mean μ and dividing by the σ. The typified normal distribution $N(0, 1)$ is called the *standard normal distribution* and it is the reference distribution for many statistical purposes.

Along with the normal distribution there are some derived distributions that are useful for inference purposes. We can define the *student's "t" distribution with $n-1$ degrees of freedom* as the distribution of the statistic

$$\frac{\bar{x} - \mu}{\frac{S_1}{\sqrt{n}}} \sim t_{n-1} \ .$$

The shape of this distribution is similar to the Gaussian, symmetric and bell-shaped, but with heavier tails and, hence, more likely to produce values far from the mean.

Another important distribution is the *chi squared distribution with n degrees of freedom*, which is the sum of n squared independent Gaussian standard distributions, written

$$\sum_{i=1}^{n} Z_i^2 \sim \chi_n^2 \ ,$$

where $Z_i \sim N(0, 1)$. This is a non-symmetric distribution which only outputs positive values. Density functions for several values of n are shown in Fig. 5.18.

Finally, a *Snedecor's F with n and m degrees of freedom $F_{n,m}$* is the quotient of two independent chi squared distributions with n and m degrees of freedom divided by their respective degrees of freedom. If $X \sim \chi_n^2$ and $Y \sim \chi_m^2$, then

$$\frac{X/n}{Y/m} \sim F(n, m) \ .$$

The reason we introduce these distributions is because they play an important role in the inference processes that we will perform later.

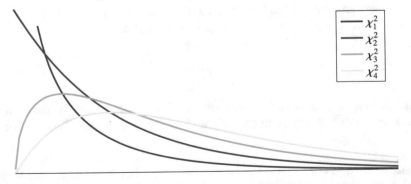

Fig. 5.18 Chi squared distributions with several degrees of freedom

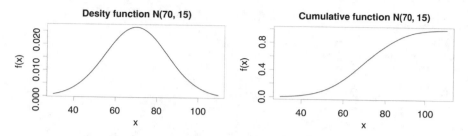

Fig. 5.19 Density and distribution functions of a normal distribution $N(70, 15)$

Suppose that we are measuring a variable, for example, the heart rate of women in the Pima tribe X and we know that this population has approximately a normal distribution with mean 70 and standard deviation 15, that is, $X \sim N(70, 15)$. We can compute the probability that a woman has a heart rate between 60 and 80 by using the cumulative distribution, and the fact that $P(60 < X < 80) = P(X < 80) - P(X < 60) = F(80) - F(60)$:

```
pnorm(80, 70, 15) - pnorm(60, 70, 15)
```

```
[1] 0.4950149
```

We can also plot its density and cumulative functions by evaluating them in a sequence of several values:

```
x <- seq(40, 100, by=0.5)
curve(dnorm(x, 70, 15), xlim=c(40, 100), col="blue", lwd=2,
      xlab="x", ylab="f(x)", main="Density function N(70,15)")
curve(pnorm(x, 70, 15), xlim=c(40, 100), col="blue", lwd=2,
      xlab="x", ylab="f(x)", main="Cumulative function N(70,15)")
```

and we obtain Fig. 5.19.

In order to empirically identify whether a given sample comes from a normal distribution we can plot a histogram and compare it with an actual normal.[30] In this example, we use `rnorm()` to simulate 10,000 values from the normal distribution $N(70, 15)$ representing the aforementioned heart rate and compare their histogram with the population distribution to check for similarities. With the following code we obtain Fig. 5.20, where we see how the simulated sample follows very closely the theoretical distribution.

```
set.seed(1)
sample <- rnorm(1000, 70, 15)
hist(sample, freq=FALSE, breaks=seq(20, 130, 10), col="seashell",
     xlab="Heart rate",
     main="Histogram for simulation of N(70,15)")
curve(dnorm(x, 70, 15), xlim=c(20, 130), col="blue",
      lwd=2, add=TRUE)
```

---

[30]In Sect. 5.2.3 we will see more accurate methods of checking it.

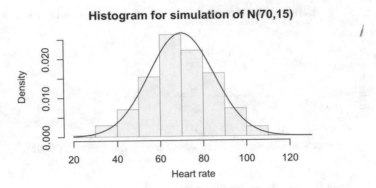

Fig. 5.20 Histogram of a simulation and density function of a normal distribution $N(70, 15)$

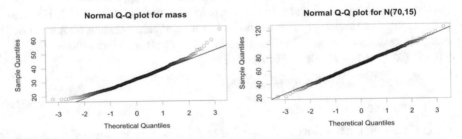

Fig. 5.21 Q-Q plots for the variable mass and a simulated sample drawn from $N(70, 15)$

Another way of comparing a given sample with a theoretical one drawn from a normal distribution is to use a Q-Q plot. These plots compare the quantiles of our distribution, this is the values separating given percentages of the individuals to study, with the ones of a theoretical distribution with the same mean a standard deviation (generally a normal distribution). The closer to the diagonal the points are, the more similar is the distribution to a normal one. We use function qqnorm() to display the Q-Q plot for the mass variable of our dataset, where each point has the quantiles of the normal reference distribution and mass as the coordinates. We also show the same Q-Q plot for a simulated random sample drawn from $N(70, 15)$, compared against the theoretical one. We see both results in Fig. 5.21. Observe how both cases fit quite well a normal distribution. However, as we saw in Fig. 5.13, the empirical distribution of mass is a Gauss bell slightly right-tailed, which implies that its upper quantiles are allocated a bit higher than the ones in a normal distribution, precisely the situation we have in Fig. 5.21, left.

```
qqnorm(mass, main="Normal Q-Q plot for mass")
qqline(mass, col="blue", lwd=2)  # adds a reference line

qqnorm(rnorm(1000, 70, 15), main="Normal Q-Q plot for N(70,15)")
qqline(rnorm(1000, 70, 15), col="blue", lwd=2)
```

The normal distribution has the so-called additive property: the sum of two independent normal distributions is a normal distribution with mean and variance given by the sum of means and variances, respectively. This does not happen in general for every distribution, but it turns out that the sum of *many* random variables approximates very well a normal distribution. This is the reason why many real distributions are normal, because they are, in fact, the sum of many others. One of the most important theorems in probability theory is the *Central Limit Theorem* (CLT in what follows) we mentioned before, which states that, under some finiteness conditions, the sum of many independent and identically distributed random variables is, indeed, a normal distribution.[31] We could formulate it in the following way. For $X_1, X_2, \ldots, X_n$ independent random variables, with n sufficiently large,[32] all sharing the same distribution with mean μ and variance σ^2, the random variable resulting of their sum $\sum_{i=1}^{n} X_i$ follows *approximately* a normal distribution of mean $n\mu$ and variance $n\sigma^2$. Moreover, the mean of the random variables $\overline{X} = \frac{1}{n} \sum_{i=1}^{n} X_i$ follows a normal distribution $N(\mu, \sigma/\sqrt{n})$ and, therefore, the standardized mean $\overline{X}^* = \frac{\overline{X} - \mu}{\sigma/\sqrt{n}}$ follows a standard normal.

Let us check how the CLT works in order to develop some intuition about it. Consider a Poisson distribution $Po(4)$ and check the CLT with it. With the following code:

```
m = 10000   # sample size of the final variable
lambda = 4   # poisson parameter
par(mfrow=c(2, 2))
for (n in c(1, 5, 10, 100)) {
  x <- NULL #initialization of x
  for (i in 1:m) {
        x <- c(x, mean(rpois(n, lambda)))
  }
  x.normalized <- (x-lambda)/(sqrt(lambda/n))
  hist(x.normalized, col="seashell",
  main=paste("Mean of", n, "Po(4)"), xlab="Value")
}
```

we can visualize in Fig. 5.22 the transformation of the average of n Poisson distributions $Po(4)$ as n increases, from a Poisson to a standardized normal. We can see how the larger the n, the more symmetric and close to a standard normal is the distribution of the samples. Moreover, we can consider that any variable in our dataset comes from a population and that sampling from this population arises also from random variables. This is the reason of using simple random sampling: it guarantees that every drawn from a sampling is actually independent from the previous ones and hence, n-sized samplings can be considered as n independent and identically distributed random variables. So, given a set of n numbers, we can conclude that the sample mean follows indeed a normal distribution by the CLT.

---

[31] Introduced by de Moivre and developed by Laplace, Lyapunov, and many others, this theorem has a long history and many proofs and versions, including some in which, under extra finiteness conditions, the variables involved do not need to be identically distributed. We refer to the interested reader to [5].

[32] This is a mathematical limit where the larger n, the better the approximation.

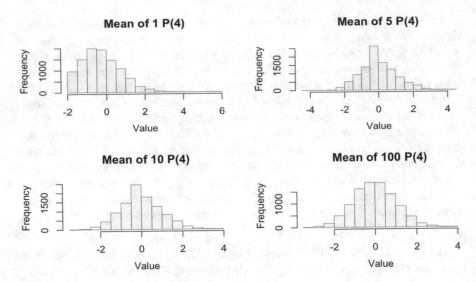

Fig. 5.22 Mean of a Poisson distribution as n grows

Indeed, for our variable `mass`, we can repeat the previous process to take samples of the standardized mean of this variable and obtain a standard normal. The following code:

```
n = 100
m = 10000
mu = mean(mass, na.rm=TRUE)
sigma = sd(mass, na.rm=TRUE)
x <- NULL
for (i in 1:m) {
  x <- c(x, mean(sample(na.omit(mass), n)))
}
y = (x-mu)/(sigma/sqrt(n))
hist(y, col="seashell",
     main=paste(paste("Mean of 100",sep=" "),"mass",sep=" "),
     xlab="Value")
```

produces the histogram in Fig. 5.23. See how close to a normal is this histogram when compared to the empirical density function in Fig. 5.13.

5.2.3 Confidence Intervals and Hypothesis Tests

5.2.3.1 Point Estimation

Once we have a simple random sample $x_1, x_2, \ldots, x_n$ from an specific population X (say, $X \sim N(\mu, \sigma^2)$), we try to guess the parameters of the population (μ and σ^2) by using *estimators* which are just functions of the sample $f(x_1, x_2, \ldots, x_n)$.

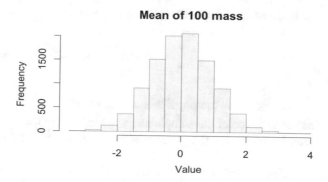

Fig. 5.23 Mean distribution of 10,000 samples of 100 individuals each from the mass variable. The Central Limit Theorem guarantees that this distribution is normal

Any function of the sample can be an estimator. All the measures introduced in Sect. 5.1 are candidates for estimators, and we will refer to them as the sample mean $\bar{x}$, the sample variance S^2, and so on. When they are computed over different samples, they can have different values as a result. Hence, they are random variables, because their values depend on the concrete sample used to compute them. As random variables, we can compute their moments, such as the expected value. We use the so-called *principle of analogy* in which the construction of an estimator is based in the moment we want to estimate. For instance, if we want to estimate the mean of a population, we shall use the sample mean.

We define some properties to decide which is the best estimator to be used in each case. In order to do that, consider an estimator $\widehat{\alpha}$ for a parameter α, and define the *mean squared error* $MSE(\widehat{\alpha}) = E[(\widehat{\alpha} - \alpha)^2]$, which can be also decomposed as

$$MSE(\widehat{\alpha}) = V(\widehat{\alpha}) + [\alpha - E(\widehat{\alpha})]^2.$$

Hence the minimization of the mean squared error depends on the variance of the estimator and the difference between the expected value of the estimator and the parameter we want to estimate, which is called the *bias*. Estimators having the desirable property of $E(\widehat{\alpha}) = \alpha$ are called *unbiased*. Therefore, we want unbiased estimators with minimum variance. Sometimes it is not possible to have both things at the same time and we seek for a trade-off between them.

It can be proved that the expected value of the sample mean $\bar{x}$ from a distribution X with mean μ is precisely $E(\bar{x}) = \mu$, hence $\bar{x}$ is unbiased for μ. This means that, if we want to estimate the mean of a population, the estimator $\bar{x}$ is unbiased and, in addition, its variance is minimum. When the latter is satisfied we call such an estimator *efficient*. We must always remember that the sample mean is not the actual mean, but it is the best approximation we can get. For example, if we want to estimate the real mean of the Pima Indian women `triceps`, we can use the sample mean, so

```
mean(triceps, na.rm=TRUE)
```

```
[1] 29.15342
```

is our guessing candidate to be the population mean. For the sake of completeness, we also include the variance of the sample mean, which is $V(\overline{x}) = \frac{\sigma^2}{n}$. Note that, the larger the sample, the smaller the variance of the mean, and hence the estimation error.

The expected value for the sample variance is $E(S^2) = \sigma^2 - \frac{\sigma^2}{n}$, where σ^2 is the population variance of X. Estimating the population variance by this number is not that good, because it underestimates the variance with a bias equal to $-\frac{\sigma^2}{n}$. This is the reason why we discussed about the existence of the unbiased variance S_1^2 in Sect. 5.1, whose expected value turns out to be $E(S_1^2) = \sigma^2$, with the desired unbiased property. Recall from Sect. 5.1 that the built-in function `var()` in **R** does not compute S^2 but S_1^2, and the **R** function `sd()` computes $\sqrt{S_1^2}$. Hence, in order to estimate the population variance of the `triceps` variable, we use

```
sd(triceps, na.rm=TRUE)^2
```

```
[1] 109.7672
```

5.2.3.2 Confidence Intervals

Achieving a perfect estimation of a parameter of a population is essentially impossible. No matter what estimator we use, the exact value of the population will not be exactly the same as the calculated from a sample. Nonetheless, we can be confident that the actual value of the parameter is going to be close to our estimation in some sense. But, how confident can we be with the data we have? Can we give a bound error for our estimation? In statistics we work with probabilities so we need to ensure that our estimations are close to the real value with some probability. How far from our estimated value for the mean of `triceps` is the actual value of the population?

We introduce a *confidence interval* $I_\lambda = (\lambda_1, \lambda_2)$ for a parameter λ with *confidence* γ as the one satisfying that the probability of having $\lambda \in I_\lambda$ is $P(\lambda_1 < \lambda < \lambda_2) = \gamma$. We call *significance level* to the number $\alpha = 1 - \gamma$. Usually, γ will be a high value (95%, 99%, etc.) in order to ensure than the confidence on the estimation is enough for our purposes. As we increase the confidence, the interval will be wider, since the necessary range to contain a growing confidence will broad out.

Confidence intervals are constructed around a central point, called pivot. For example, we want to find a confidence interval for the population mean μ following a normal distribution $N(\mu, \sigma)$, by using a simple random sample $x_1, x_2, \ldots, x_n$

with confidence of 0.95. The point estimator for μ is the sample mean $\bar{x}$, so we will center our interval on $\bar{x}$ and we use the pivot $p = \frac{\bar{x}-\mu}{\sigma/\sqrt{n}}$ which follows a normal $N(0, 1)$. Thanks to this, we can use the symmetry of the normal distribution around zero to compute a value z such that $P(-z < p < z) = 0.95$, that is, satisfying that $P(p < z) = 0.975$ and then, from the last formula we can operate and isolate μ in the center of the inequalities, obtaining that

$$P(\bar{x} - z\frac{\sigma}{\sqrt{n}} < \mu < \bar{x} + z\frac{\sigma}{\sqrt{n}}) = 0.95.$$

Hence, we have an interval containing μ with a probability of 0.95. To compute an interval in **R** for the population mean of the `triceps`, we have to first suppose that it is a normal distribution (which is usually the case given the CLT) and then construct it manually. The problem here is that we *do not know* the population variance, as it is going to be almost always the case.

In order to avoid this problem we use the Student's "t" distribution with $n - 1$ degrees of freedom, which is the one followed by the pivot $p = \frac{\bar{x}-\mu}{S_1/\sqrt{n}}$, including $\bar{x}$ and μ again. Exactly in the same way as we did before, we derive an interval for the mean μ, obtaining

$$P\left(\bar{x} - t_{n-1}\frac{S_1}{\sqrt{n}} < \mu < \bar{x} + t_{n-1}\frac{S_1}{\sqrt{n}}\right) = 0.95,$$

where t_{n-1} is a value satisfying $P(-t_{n-1} < p < t_{n-1}) = 0.95$. Now, we can compute the interval for the variable `triceps` as follows:

```
conf.level <- 0.95
t <- qt((1 - conf.level) / 2, df=length(triceps) - 1,
        lower.tail = FALSE)
sigma <- sd(triceps, na.rm=TRUE)
mean <- mean(triceps, na.rm=TRUE)
sd.mean <- sigma / sqrt(length(triceps) - sum(is.na(triceps)))
interval <- c(mean - t * sd.mean, mean + t * sd.mean)
interval
```

```
[1] 28.26918 30.03766
```

This means that the true value for the mean of the `triceps` variable for the Pima women is between 28.26918 and 30.03766 with a probability of 0.95.

We can also compute the same interval for the mean using function `MeanCI()` from the `DescTools` package, already mentioned in Sect. 5.1.1 (when computing the mode).

```
MeanCI(triceps, na.rm=TRUE, conf.level=0.95)
```

```
    mean    lwr.ci    upr.ci
29.15342 28.26859 30.03825
```

There is one more way to compute this interval, from a more modern and computational approach, called *Bootstrapping*. This is a resampling method that consists on sampling many times from the original data with replacement, and then computing estimators and confidence intervals from these samplings. Bootstrap advantage (as well as other resampling methods) is that we generate many samples and we are able to estimate, in some sense, the variability of our estimators with only one sample. For more about Bootstrapping and other resampling methods, we refer to the reader to [2] and [11]. We resample 10,000 times with the function boot () applied to the variable triceps without NAs, and the estimator we use, the mean. Then we compute the intervals with boot.ci (), both in the boot package:

```
library(boot)
set.seed(1)
resampling <- boot(na.omit(triceps), function(x, i) mean(x[i]),
                   R=10000)
boot.ci(resampling, conf = 0.95)
```

```
bootstrap variances needed for studentized intervals
BOOTSTRAP CONFIDENCE INTERVAL CALCULATIONS
Based on 10000 bootstrap replicates

CALL :
boot.ci(boot.out = resampling, conf = 0.95)

Intervals :
Level        Normal              Basic
95%     (28.27, 30.03 )     (28.27, 30.02 )

Level        Percentile           BCa
95%     (28.28, 30.04 )     (28.29, 30.04 )
Calculations and Intervals on Original Scale
```

Note that similar results arise from all methods. Bootstrapping is not really making a difference in this computation but resampling methods have been demonstrated to be very powerful in many situations.

To compute an interval for the variance σ^2 from a normal population $N(\mu, \sigma)$, we can use the fact that the pivot $p = \frac{nS^2}{\sigma^2}$ has a χ^2 distribution with $n - 1$ degrees of freedom. We can proceed as before to derive an 0.95 confidence interval for σ, resulting on

$$P\left(\frac{nS^2}{\chi_2} < \sigma^2 < \frac{nS^2}{\chi_1}\right) = 0.95,$$

where χ_1 and χ_2 satisfy $P(\chi_1 < p < \chi_2) = 0.95$. We can use the VarCI () function (in DescTools) to compute a confidence interval for the variance of mass:

```
VarCI(mass, na.rm=TRUE, conf.level=0.95)

     var     lwr.ci     upr.ci
47.95546  43.46581   53.18227
```

In general, we can compute intervals for any parameter as long as we have a pivot satisfying the properties quoted above. In each case, we can derive formulae for these intervals and implement them in **R** manually. Nevertheless, the next section is devoted to hypothesis tests and it is often the case that functions in **R** which perform these tests, and also compute the confidence intervals.

5.2.3.3 Hypothesis Testing

Hypothesis testing is a way to check guesses we make about our distributions. With the same starting point as with confidence intervals, we adopt a different point of view. Now, we do not want just a set of possible values for one parameter, instead, we test if a statement is reliable based on the sample we are provided with. We can ask different questions about a distribution. For example, in our Pima Indian dataset, we wonder whether our variables are normally distributed, or if it is admissible to assume that the mean of `pressure` for the whole population is 70 knowing that the sample mean is 72.40518. We test claims considering the sample data as the only information we have for the actual population. Note that if we want to perform a test about the population mean, we have at hand more information than just the mean in order to determine if our assumptions are reliable or not.

The way we test these claims is by formulating hypothesis. The *null hypothesis*, or H_0, is a statement that is not rejected unless proven otherwise. As an opposite to the null hypothesis, the test provides the *alternative hypothesis* denoted by H_1, which may include situations that H_0 does not account for. With two possible alternatives, H_0 and H_1, we have four possible options, depicted in the following table:

	H_0 true	H_0 not true
Do not reject H_0	✓	ε_2
Reject H_0	ε_1	✓

If the real scenario is that H_0 is true and the test does not reject it, we are committing no mistake. The same happens if the reality is that H_0 is not true, and the test shows enough evidence to prove that it had to be rejected. However, there are two possible ways of making a mistake. We could reject the null hypothesis while it is in fact true (type 1 error) or we could not reject the null hypothesis being false (type 2 error). We call ε_1 and ε_2 the probabilities of committing each mistake, respectively. In order to derive the test, we must decide which is the main probability to minimize. Here, we have to recall the asymmetry of our hypothesis: we are testing whether H_0 can be rejected or not, not if we accept H_0 or H_1. The question we are really wondering reads: Statistically speaking, is there enough evidence in our data to reject the null hypothesis or not?

We should think about the null hypothesis as innocence in a judgment process. The presumption of innocence is a legal principle in which somebody is considered innocent unless proven guilty. Here we consider that H_0 is true unless proven false. We prefer to minimize the probability of sentencing an innocent person by reducing ε_1. However, this may increase the probability of committing the type 2 error, ε_2, not sentencing a guilty person. Depending on the case we must decide which error we do prefer to minimize. Usually, especially in natural sciences, we prefer to minimize ε_1, and we fix it to be a very small value, 0.05 or 0.01. We also call this quantity the *significance level*. The significance level is set in advance depending on the probability of being wrong that the researcher is willing to assume.

This way we can construct our tests. Once we fix the value ε_1, we need to calculate the *test statistic* T. This is a function of the data sample, just as the estimators from the previous sections. Given this test statistic we will reject H_0 if T belongs to the so-called *critical region*, a region of values too extreme to be supported by the statistical evidence. The way of computing critical regions is very similar, in spirit, to the construction of confidence intervals, with a pivot relating the terms of our guessing and with a known distribution. We will not pursue into the details here.

The critical regions grow with ε_1, being bigger when ε_1 is also bigger, and increasing the probability of rejecting the null hypothesis. Therefore, we will incorrectly reject the null hypothesis with a probability $\varepsilon_1 = 0.05$, or 5 out of 100 innocent people will be considered guilty. If, instead, we choose $\varepsilon_1 = 0.01$ we will reduce the number of mistakes. It may seem a good idea to reduce ε_1 as much as possible, but as we said, this will increase ε_2, the probability of considering innocent a guilty person.

Computing critical regions for a particular value of ε_1 is a mathematical task that can be done easily for some well known distributions as the normal. But it turns out that we could be not rejecting a null hypothesis with a significance level of $\varepsilon_1 = 0.05$ but it would be rejected with $\varepsilon_1 = 0.06$. In these cases, we might be interested in the limit of rejection, that is, the minimum value for the significance level in which we reject the null hypothesis (or the maximum in which we do not). This is the so-called *p-value*: for significance levels smaller than it, we do not reject the null hypothesis while it is rejected for greater ones. We consider that there is enough statistical evidence to reject the null hypothesis, when the probability of being wrong is greater than the *p*-value. This number is more complicated to compute than performing a single test with a given significance level but it is easily calculated with a computer and it is widely used, since it informs us, at once, which significance levels reject the null hypothesis and which do not.

In terms of the critical region the *p*-value is the probability that, when the null hypothesis is true, the test statistic would be greater than or equal to the actual observed result T. This might sound a little cryptic, but the idea is that the *p*-value is the probability of obtaining a T value even bigger than the actually observed. If this probability is small it is because T is inside the critical region and we must reject H_0, and the opposite otherwise. For example, if the *p*-value is 0.034 we know that we do not reject the null hypothesis for a critical region containing the 0.025

probability mass, but we will for a larger critical region, of a 0.05 for example. The size of the critical region will depend on the value of ε_1: the larger the error, the bigger the critical region. As a general rule, high values of the p-value do not reject the null hypothesis since a small ε_1 will create a small critical region.

There are, mainly, two types of hypothesis tests: *Parametric and non-parametric.* Former case are tests for parameters of a given population. For example, a test for the mean μ of a normal distribution $N(\mu, \sigma)$. The latter are tests in which we check a global property of a distribution: if it is normally distributed, whether it is random or not, and so on.

5.2.3.4 Parametric Tests

The parametric tests have also several different types. For all the tests used here, we assume that our population follows a normal distribution.[33] Assume that we want to test about a parameter λ from a fixed distribution, the null hypothesis being $H_0 : \lambda = \lambda_0$, where λ_0 is a concrete value for it. There are three possible alternatives for H_1:

- $H_1 : \lambda = \lambda_1$, where λ_1 is a different concrete value for the parameter. These tests are called *Neyman–Pearson* tests.
- $H_1 : \lambda \neq \lambda_0$, the *two-sided, two-tailed,* or *bilateral* tests, because the critical region will consist of two disconnected intervals.
- $H_1 : \lambda < \lambda_0$ or $H_1 : \lambda > \lambda_0$, the *one-sided, one-tailed,* or *unilateral* tests, because the critical region will consist only on one interval.

The difference between these tests is the construction of the critical region. In the case of a Neyman–Pearson test, the critical region will be computed in the same way as in a unilateral test, the inequality indicating the alternative hypothesis coherent with the number λ_1. That is, if λ_1 is greater than λ_0, then the alternative hypothesis is $H_1 : \lambda_1 > \lambda_0$. The concrete value λ_1 can be used to compute the probability ε_2, whose complementary $1 - \varepsilon_2$ is called the *power test*, and measures the probability of correctly rejecting H_0 when it is false.

More general tests can be considered and transformed into the ones listed above. For instance, to test the null hypothesis that the population means of two normal distributions are equal, that is, $H_0 : \mu_1 = \mu_2$, we could just define a new variable as the difference of them and then test for the difference of their means to take a particular value, for instance, $H_0 : \mu_1 - \mu_2 = 0$. Note that we have to decide which of these tests make sense under each circumstance. For example, suppose we are performing a test about the value of the mean α of a non-negative variable, and we want to check the null hypothesis $H_0 : \alpha = 0$. Among the alternatives, $H_1 : \alpha < 0$ or $H_1 : \alpha \neq 0$ do not make any sense, even though the test can be calculated for

---

[33]It is not a trivial assumption to do, but we will stick to this, which covers so many real cases, to illustrate the tests.

them. It is our mission to select the correct alternative hypothesis. Note also that, when performing a hypothesis testing, we prefer to use a unilateral contrast in case we really know that it is the one making sense, because the power of the contrast is going to focus into that side of the value for the null hypothesis.

We now perform hypothesis testing in **R**. We were wondering if the variable pressure could have mean 70, with 100 or 71 as alternatives, that is, we perform a hypothesis test for the mean of this variable, where $H_0 : \mu = 70$ and $H_1 : \mu > 70$. We check it with the function t.test():[34]

```
t.test(pressure, mu=70, alternative="greater")
```

```
        One Sample t-test

data:  pressure
t = 5.259, df = 732, p-value = 9.516e-08
alternative hypothesis: true mean is greater than 70
95 percent confidence interval:
 71.65196        Inf
sample estimates:
mean of x
 72.40518
```

We obtain a very instructive response. The t-test statistic is 5.259 and the p-value (i.e., the minimal value to reject the null hypothesis) is very small, and hence we can reject it at almost every significance level with a very small probability of being wrong. Indeed, the value 70 is smaller compared to the sample mean of pressure which is 72.4. Hence, with our data, we must reject the possibility of the population mean being only 70. We can try with a closer number, such as 72 and, in this case, against the alternative of being just different.

```
t.test(pressure, mu=72, alternative="two.sided")
```

```
        One Sample t-test

data:  pressure
t = 0.88595, df = 732, p-value = 0.3759
alternative hypothesis: true mean is not equal to 72
95 percent confidence interval:
 71.50732 73.30305
sample estimates:
mean of x
 72.40518
```

Now t is a less extreme value (closer to 0) and, as a consequence, p-value is somewhat large. It is greater than both 0.01 and 0.05 and hence we cannot reject the null hypothesis for these significance levels. We have no evidence to reject the fact that the mean for pressure can be 72, when using our data. We can also see

---

[34]The name is because being the variance unknown the underlying distribution used to perform the test is the student's "t".

the 95% confidence interval for the mean, as we said in the previous section. For the sake of completeness, we make explicit the interpretation for a more conflictive value for H_0.

```
t.test(pressure, mu=71.4, alternative="two.sided")
```

```
        One Sample t-test

data:  pressure
t = 2.1979, df = 732, p-value = 0.02827
alternative hypothesis: true mean is not equal to 71.4
95 percent confidence interval:
 71.50732 73.30305
sample estimates:
mean of x
 72.40518
```

In this case, we obtain a p-value which is between 0.01 and 0.05! What is the interpretation here? We can reject the null hypothesis at a 0.05 level but not at 0.01. That is, if we minimize the probability of being wrong when rejecting it to a 1%, we just cannot reject, while allowing a 5% leaves us rejecting it safely. Once a t.test() is performed, we can access to its internal calculations. For example, we can type

```
test <- t.test(pressure, mu=70, alternative="greater")
test$p.value
```

```
[1] 9.515989e-08
```

to obtain the concrete p-value.

With this test, we can also compare means of two populations. For example, perform a test for the difference of population means for insulin and glucose. The null hypothesis is then $H_0 : \mu_1 - \mu_2 = 0$ the alternative being that it is not equal (or the means are different).

```
t.test(insulin, glucose, mu=0, alternative="two.sided")
```

```
        Welch Two Sample t-test

data:  insulin and glucose
t = 5.5647, df = 420.03, p-value = 4.688e-08
alternative hypothesis: true difference in means is not equal
to 0
95 percent confidence interval:
 21.90042 45.82250
sample estimates:
mean of x mean of y
 155.5482  121.6868
```

We clearly reject this hypothesis, in view of the p-value. Also, just by comparing the sample means, they are so different that clearly do not represent the same population means. Hence, we perform a one-sided test, to check, as a null hypothesis, if the population mean of the first variable is greater than the second in view of our data, the alternative being the opposite.

```
t.test(insulin, glucose, mu=0, alternative="less")
```

```
        Welch Two Sample t-test

data:  insulin and glucose
t = 5.5647, df = 420.03, p-value = 1
alternative hypothesis: true difference in means is less than 0
95 percent confidence interval:
     -Inf 43.89268
sample estimates:
mean of x mean of y
 155.5482  121.6868
```

Here, we do not reject that assumption, as it was expected. Moreover, we can check if the difference between the means is around 30.

```
t.test(insulin, glucose, mu=30, alternative="two.sided")
```

```
        Welch Two Sample t-test

data:  insulin and glucose
t = 0.63458, df = 420.03, p-value = 0.5261
alternative hypothesis: true difference in means is
not equal to 30
95 percent confidence interval:
 21.90042 45.82250
sample estimates:
mean of x mean of y
 155.5482  121.6868
```

obtaining that we neither reject this null hypothesis.

With `VarTest()` in package `DescTools`, we can perform tests and compute confidence intervals for the variance. For example, the following test checks whether the variance of `pressure` is 153, not founding enough statistical evidence (at a 95% confidence level) to reject it:

```
VarTest(pressure, sigma.squared=153)
```

```
        One Sample Chi-Square test on variance

data:  pressure
X-squared = 733.52, df = 732, p-value = 0.9544
alternative hypothesis: true variance is not equal to 153
95 percent confidence interval:
 138.7477 170.3223
sample estimates:
variance of x
     153.3178
```

We can also perform a test to check out the ratio between two variances with var.test from stats. Let us test the null hypothesis $H_0 : \frac{\sigma_1^2}{\sigma_2^2} = 3$ against the quotient being different from 3,

```
var.test(pressure, mass, ratio=3, alternative= "two.sided")

        F test to compare two variances

data:   pressure and mass
F = 1.0657, num df = 732, denom df = 756, p-value = 0.3855
alternative hypothesis: true ratio of variances is not equal to 3
95 percent confidence interval:
 2.768995 3.691989
sample estimates:
ratio of variances
        3.197088
```

and do not reject this hypothesis, as we could suspect given that the sample ratio is 3.2. The name F for the test and the test statistic comes from the Snedecor's F distribution, introduced in Sect. 5.2.2.

5.2.3.5 Non-parametric Tests

We now present some non-parametric tests which allow us to test some more general distribution features and how to perform them in **R**. The *runs test* for a set of ordered numbers (for example, in time series) with two possible values, say 1 and 0, tests the null hypothesis that the number of runs, defined as sequences of consecutive 1s or 0s in our series, is arbitrary or erratic or follows a pattern. This serves to detect if the sample is random or not. If this number of runs, R, is too small or too large, the numbers are not random. For example, if we toss a coin, with 1 for head and 0 for tail, and we obtain large sequences with the same number, we can assume that the coin is loaded on one of the sides. Moreover, if we obtain series with a clear pattern, such as 1,0,1,0,1,0,1,0,1,... we could also suspect that the tossing is not random: the following result can be predicted by the previous one.

In general, the number of runs R follows a distribution which, for large samples, is a normal distribution. To perform it, divide the series into two subsets of 1s and 0s; if the series has more than two values, they are divided into two categories, the ones smaller than the median and the ones which are greater. To perform this test in **R** we use the function runs.test() in package snpar.

As an example, let us test the randomness of some distributions. For the following sequence of ten coin tosses:

```
library(snpar)
runs.test(c(1, 1, 1, 1, 1, 1, 1, 1, 0, 0))

        Approximate runs rest

data:   c(1, 1, 1, 1, 1, 1, 1, 1, 0, 0)
```

```
Runs = 2, p-value = 0.01287
alternative hypothesis: two.sided
```

we obtain that the null hypothesis of randomness is rejected with a significance level of 0.05 (but not with a 0.01 level). This means that this small sample can be considered non-random with a 95% confidence. Consider now a randomly generated set of 10^5 numbers sampled from a standard normal and take as 0 the negative values and 1 the positive ones.

```
set.seed(5)
sample.vector <- rnorm(100000, 0, 1)
binary.vector <- ifelse(sample.vector < 0, 0, 1)
runs.test(binary.vector)

        Approximate runs rest

data:  binary.vector
Runs = 50172, p-value = 0.268
alternative hypothesis: two.sided
```

As we could suspect drawing the sample randomly, we do not reject the null hypothesis of randomness.[35]

The *Kolmogorov–Smirnov* test is used to test whether two samples are drawn from the same distribution. Imagine we want to check the reliability of our collected data glucose and diabetes coming from the same distribution. We test as null hypothesis the fact that they do, with the ks.test:

```
ks.test(glucose, pressure)

p-value will be approximate in the presence of ties
Two-sample Kolmogorov-Smirnov test

data:  glucose and pressure
D = 0.80399, p-value < 2.2e-16
alternative hypothesis: two-sided
```

The small p-value makes us reject the null hypothesis; therefore, the samples do not come from the same distribution. As another example, test two randomly generated samples drawn from a Poisson distribution of parameter 5:

```
set.seed(5)
ks.test(rpois(100, 5), rpois(200, 5))

p-value will be approximate in the presence of ties
Two-sample Kolmogorov-Smirnov test

data:  rpois(100, 5) and rpois(200, 5)
D = 0.07, p-value = 0.8996
alternative hypothesis: two-sided
```

---

[35]Running the test several times, for different samples from the normal distribution, might lead in some cases (very few, but some) to reject the randomness hypothesis. Try to check for set.seed(32).

In this case we obtain that, indeed, both come from a Poisson distribution.

As we have previously explained, many situations in data analysis involve normal populations, and variables are assumed to be normal. The *Shapiro–Wilk test* is used to test if a sample is likely to have been drawn from a normal distribution. This is a very important condition since every test we performed for the population mean, for example, is only reliable if the population is a normal distribution. For example, let us test if the data of the variable `insulin` comes from a normal distribution.

```
shapiro.test(insulin)

        Shapiro-Wilk normality test

data:   insulin
W = 0.8041, p-value < 2.2e-16
```

Hence, we reject the null hypothesis of normality. As mentioned, being drawn from a normal distribution is a very strong condition to satisfy. Even being a necessary condition to do many of the inference methods introduced here, sometimes we do not have this condition and our analysis is less efficient. We can check how this test really captures the normality of a sample, randomly generated from a normal distribution.

```
set.seed(5)
shapiro.test(rnorm(5000, 0, 1))

        Shapiro-Wilk normality test

data:   rnorm(5000, 0, 1)
W = 0.99961, p-value = 0.4446
```

Hence, of course, the sample corresponds to a variable normally distributed.

5.2.4 Exercises

Exercise 5.17 Consider that 10 friends meet at a party. One asks "What is the probability that, at least, the birthday of two of us occurs the same day?" Compute this probability by using combinatorics. Define a function such that, given a number of people n in the party, computes the probability that, at least, two of them have the same birthday. Store the corresponding probability for the first 50 numbers in a vector, and plot it to observe its growth rate.

Exercise 5.18 Consider a French deck of cards, and draw five cards randomly with and without replacement. A poker game between four people starts by drawing five cards for each one, with one possible discard and replacement of one card for each player. Simulate this in **R**.

Exercise 5.19 Let X be the random variable consisting of the sum of the numbers of a dice when we toss it two times. Plot the mass and distribution functions for the variable X and compute the probability of obtaining an even and an odd number.

Exercise 5.20 A company applies, independently, for 10 grants and the probability of succeeding is always 0.15. What is the probability of obtaining at least 4 grants, more than 2 grants, and exactly 3 grants? What is the maximum number of grants that the company can get with a probability of 0.80?

Exercise 5.21 Compute confidence intervals of 99% for the arithmetic mean of the quantitative variables in the Pima Indian dataset and compare their lengths.

Exercise 5.22 Test whether the difference of means of the variables insulin and mass (from Pima Indian dataset) is greater than zero with a significance level of 0.05.

Exercise 5.23 Perform a hypothesis test to decide which variance, for the variables mass and triceps from Pima Indian dataset, is smaller. Hint: test whether the quotient between the variances is smaller than 1.

Exercise 5.24 Check if the some of the variables of the Pima Indian dataset comes from a Poisson distribution. To select the Poisson parameter to be compared with, use the mean of the variable in turn.

5.3   Multivariate Statistics

So far we have investigated, measured, or tested the variables in our dataset separately, analyzing their properties and comparing them. However, we have disregarded possible relations between those variables, one of the main parts in data analysis if not the most important. The search for such relationships is fundamental in order to make conclusions about the dataset at hand and derive a global picture of its structure and implications between variables. Studying the distribution of a variable by means of the measures of Sect. 5.1, or inferring conclusions with the tools in Sect. 5.2 is undoubtedly valuable and still we are missing an important part of the analysis. Go back to Fig. 4.30 where a relationship between two variables of the iris dataset is hinted and moreover drawn. Many questions arise: what defines a relationship? What kind of relationships are there? How can we check and assess mathematically (and not just because of the intuition of a plot) that two or more variables are related? Needless to say that understanding this topic is crucial for so many applications. Finding a relationship between the use of a medication and the decline of an illness will prove a drug valuable, and learning that the consumption of some product is related with the sales of another one might rethink completely a marketing strategy.

This section starts by addressing the analysis related to quantitative variables and, after this, studies categorical variables.

5.3.1 Correlation and Bivariate Statistics

In a dataset containing several different variables, we start by studying relationships with a pairwise approach. Two variables are related if the trend of one implies the trend of the other, for example, as one grows, the other one does it as well. We call this relation between two variables *correlation*. These relations can lead us to interpret and to provide real significance between the variables. Nevertheless, statistical *correlation does not necessarily imply causality*. When dealing with variables in the real world, conclusions must be driven carefully and checked with more samples since, sometimes, the numbers can show relationships among variables which do not make sense in real life, a phenomenon called *spurious correlation*. It is also important to note that the behavior of one variable might imply the trend on another one, but not backwards. For example, the salary implies the expenses in leisure, the more salary the more expenditure in leisure, but the counterpart is not true, it does not matter how much we spend this weekend, we will not get a salary rise because of it.

In the `PimaIndiansDiabetes2` dataset, we begin by plotting `mass` with `pressure` and `triceps` by

```
ggplot(PimaIndiansDiabetes2, aes(x=mass, y=pressure)) +
  geom_point()
ggplot(PimaIndiansDiabetes2, aes(x=mass, y=triceps)) +
  geom_point()
```

obtaining the graphs of Fig. 5.24.

We observe in the plots that there seems to be no relation between `mass` and `pressure` as the points are distributed in a random fashion with no particular recognizable pattern (apart from the fact that the shape of the whole plot is somewhat circular). However, in the second plot, we can appreciate that there is some *positive relationship* between `mass` and `triceps`. This positive relation means that as the values of `mass` increase, so do the values of `triceps`. Of course, this relationship is not true for all observations, since there are points with a high value for the first variable and a small one for the second, but the trend is that both increase together,

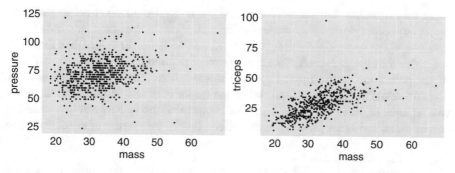

Fig. 5.24 Scatterplots of `pressure` vs. `mass` (left) and `triceps` vs. `mass` (right)

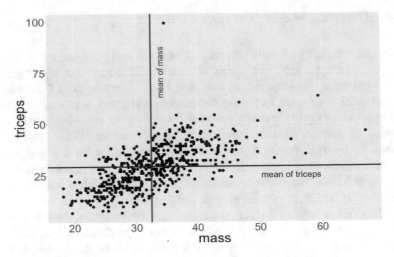

Fig. 5.25 Scatterplots of mass (left) and triceps (right) with their means

more or less, in the same proportion. We say that this relation is linear since it seems that the joint trend can be modelled by a line. That is, the common increase is related by a line, with some error.

In order to measure the linear relationship between variables let us plot the scatterplot of mass and triceps together with their arithmetic means as in Fig. 5.25, with the following code:[36]

```
ggplot(PimaIndiansDiabetes2, aes(x=mass, y=triceps)) +
  geom_point() +
  geom_vline(xintercept=mean(mass, na.rm=TRUE), colour="red",
             size=1.2) +
  geom_hline(yintercept=mean(triceps, na.rm=TRUE), colour="blue",
             size=1.2) +
  annotate("text", label="mean of mass", angle=90, x=34, y=75,
           size=4, colour="red") +
  annotate("text", label="mean of triceps", x=55, y=25, size=4,
           colour="blue")
```

In this situation, the scatterplot is divided into four different rectangles. The upper left rectangle contains women with smaller mass than the mean but greater triceps, the upper right one stands for women with greater values than the mean for both variables, and so on. Note that a positive relation is given when the majority of the points are in the upper-right and lower-left rectangles, as in this case. A negative relationship would occur when the majority of the points are in the other two regions. In some sense, the degree of linear relationship between two variables can be measured as the joint variance of them, this is, the average of the distances to

---

[36]Here we use the function annotate from ggplot2 to insert text in the figure, in the indicated coordinates.

the means of the variables. This is exactly what is measured for two variables with values (x_1, y_1), (x_2, y_2), ..., (x_n, y_n) using the *covariance*,

$$S_{xy} = \frac{\sum_i (x_i - \overline{x})(y_i - \overline{y})}{n}.$$

When the variables are positively related, the covariance will be positive as well, whereas it will be negative when they present a negative relationship. Also note that the covariance of a variable with itself recovers the variance of the variable: $S_{xx} = S_x^2$.

Similarly to what happened in Sect. 5.1 with the variance, **R** implements the *unbiased covariance* which is the same number but replacing the denominator by $n - 1$ instead of n, because it has better inference properties (as seen in Sect. 5.2) and hence it is more widely used. In order to compute it, we use the function `cov()` from the `stats` library which is included by default.

```
cov(mass, pressure, use="pairwise.complete.obs")
```

```
[1] 24.64499
```

```
cov(mass, triceps, use="pairwise.complete.obs")
```

```
[1] 46.72566
```

Note that the argument `use` determines what should be done with the missing observations. The problem is not so easy as the case of one variable since with two variables there could be NAs in both variables but there can be observations with NA for the first but not for the second, the other way around, or NA for both variables. The possible values for the argument `use` are `"everything"`, `"all.obs"`, and `"complete.obs"`.[37]

The covariance measures the common variance, hence, the greater the variance of each variable, the greater the covariance. This makes the covariance a deceiving tool for measuring relation, since it depends on the variances of each variable separately. For example, by computing the covariance we do not know if `mass` has a greater linear positive relation with `pressure` or with `triceps`, as the obtained numbers are in different scales and thus cannot be compared. In order to deal with this, we normalize this measure to obtain a truly informative object, the *Pearson coefficient of linear correlation* or simply coefficient of correlation, a real number in $[-1, 1]$ defined as[38]

$$r = \frac{S_{xy}}{S_x S_y}.$$

---

[37] The argument `method` can be used to compute other rare measures for the relationship between the two variables, although the default method is `"Pearson."`

[38] Note that this coefficient can be defined using the unbiased variances and covariance to obtain the same result, since the denominators cancel, either being n or $n - 1$.

If the correlation between two variables is close to 1 they have a positive linear relationship, whereas they are negatively related for values close to −1. If there is no linear relationship between the variables, the value of the covariance will be closer to zero. However, two variables can be related and yet have a null covariance (having non-linear relations). This is a very important fact and the correct sentence to say is "if two variables have close to one correlation they are positively related" and "if two variables are not related their correlation is zero"; but it is incorrect to say that "if the correlation is zero they are not related." And even for the first two sentences we are talking about a statistical relationship that might not make sense in real life (as with the case of expending more on the weekend to get a salary rise).

It should also be said that relations between variables are not always linear. Variables with strong correlations may follow non-linear relations, such as quadratic or logarithmic, modelling by a linear relation being just a poor approach.

Correlation is computed in **R** with the function cor():

```
cor(mass, pressure, use="pairwise.complete.obs")
```

```
[1] 0.2892303
```

```
cor(mass, triceps, use="pairwise.complete.obs")
```

```
[1] 0.6482139
```

Now we can compare correlations and decide which pairs of variables have a greater linear relationship. We can observe that mass and pressure have a weaker positive linear relationship than mass and triceps. We also got that the covariance of the first pair was smaller than the one of the second, but this is not always the case because correlation also depends on each variance. For example, mass and triceps have a smaller covariance but a greater correlation than insulin and pressure because the variances of the second pair are higher and, hence, when the correlation quotient is made, the result is smaller.

Computing all pairwise linear correlations of a dataset at once can be easily done in **R** by applying the same command to the whole dataset. The command cov() can be used in the same fashion (see Exercise 5.26 to explore the use of the covariance). Note that we exclude the non-numeric variables because this coefficient is not defined for factors.

```
cor(PimaIndiansDiabetes2[, -9], use="pairwise.complete.obs")
```

```
              pregnant     glucose      pressure      triceps
pregnant    1.00000000   0.1281346   0.214178483   0.1002391
glucose     0.12813455   1.0000000   0.223191778   0.2280432
pressure    0.21417848   0.2231918   1.000000000   0.2268391
triceps     0.10023907   0.2280432   0.226839067   1.0000000
insulin     0.08217103   0.5811862   0.098272299   0.1848884
mass        0.02171892   0.2327705   0.289230340   0.6482139
pedigree   -0.03352267   0.1372457  -0.002804527   0.1150164
age         0.54434123   0.2671356   0.330107425   0.1668158
              insulin        mass      pedigree          age
```

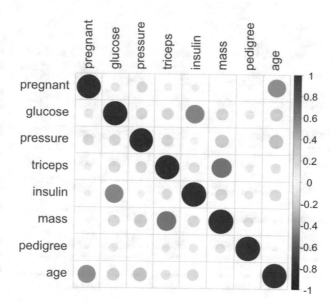

Fig. 5.26 Correlation plots for Pima Indian variables

```
pregnant 0.08217103 0.02171892 -0.033522673 0.54434123
glucose  0.58118621 0.23277051  0.137245741 0.26713555
pressure 0.09827230 0.28923034 -0.002804527 0.33010743
triceps  0.18488842 0.64821394  0.115016426 0.16681577
insulin  1.00000000 0.22805016  0.130395072 0.22026068
mass     0.22805016 1.00000000  0.155381746 0.02584146
pedigree 0.13039507 0.15538175  1.000000000 0.03356131
age      0.22026068 0.02584146  0.033561312 1.00000000
```

This code returns a matrix with all the variables as rows and columns with the corresponding correlation coefficient in each position. Of course, the correlation of one variable with itself is 1, as shown. Moreover, the matrix is symmetric because correlation does not depend on the order of the variables.

Pairwise correlations can be visualized in a friendlier representation by using the `corrplot()` function in the package with the same name, as in Fig. 5.26.

```
library(corrplot)
corrplot(cor(PimaIndiansDiabetes2[,-9],
         use="pairwise.complete.obs"))
```

The relation can also be easily visualized using `plot()` to obtain all pairwise scatterplots, as in Fig. 5.27. Even though the correlation is not quantified here, nonlinear patterns can be detected at a glance.

```
plot(PimaIndiansDiabetes2)
```

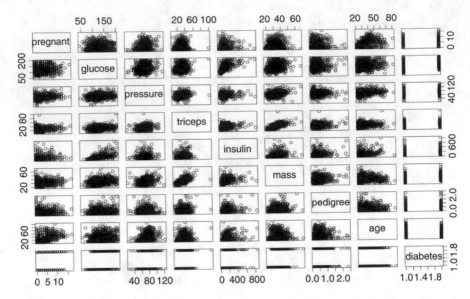

Fig. 5.27 Scatterplots for every variable in Pima Indian dataset

5.3.2 Linear Model

Once we know the correlation among pairs of variables, we compute the *regression line* for each pair. This is the line that best models the relationship between the variables, also called line of best fit. For example, in the case of triceps and mass we want to derive a linear function that models the relationship that we hinted in Fig. 5.25. That is, we want to find the numbers β_0 and β_1, such that the line

$$\mathtt{triceps} = \beta_0 + \beta_1 \mathtt{mass}$$

is the most reliable linear relation according to the dataset. In order to derive it, imagine we already know the values of this formula, and hence we can compute or *predict* values for triceps using values for mass and the formula given. If we use the data we already have for mass, we are going to obtain new numbers for triceps which usually will not coincide with the actual ones. We write (the vector of) these predicted values as $\widehat{\mathtt{triceps}}$. Our predicted values are different from the original and their differences are called the (vector of) residuals $\varepsilon = \mathtt{triceps} - \widehat{\mathtt{triceps}}$. To compute the regression line, we solve the optimization problem of finding the values of β_0 and β_1 that minimizes the sum of the squares of the residuals.[39] We call $X = \mathtt{mass}$ and $Y = \mathtt{triceps}$ and we compute the

---

[39]We minimize the squares of the residuals instead of simply the sum of them, since the predicted values can be greater or smaller than the actual values, resulting in cancellations.

regression line $Y = \beta_0 + \beta_1 X$ minimizing $\varepsilon = Y - \hat{Y}$. The answer is provided by the solution to an optimization problem, and is given by

$$\beta_0 = \overline{Y} - \beta_1 \overline{X}, \quad \beta_1 = \frac{S_{xy}}{S_x}.$$

With these values, the line $Y = \beta_0 + \beta_1 X$ is the one minimizing the residuals and this estimation method is usually called *ordinary least squares (OLS)*. We can compute the regression line very easily with the function lm, from linear model.

```
model <- lm(triceps ~ mass, data=PimaIndiansDiabetes2)
model$coefficients
```

```
(Intercept)         mass
   -3.3735       0.9895
```

Hence, the regression line is triceps $= -3.3735 + 0.9895 \cdot$ mass. We can use the summary() function to obtain more details.

```
summary(model)
```

```
Call:
lm(formula = triceps ~ mass, data = PimaIndiansDiabetes2)

Residuals:
    Min       1Q   Median       3Q      Max
-19.764   -5.068   -0.612    5.021   68.038

Coefficients:
             Estimate Std. Error t value Pr(>|t|)
(Intercept) -3.37349    1.68557  -2.001   0.0459 *
mass         0.98948    0.05016  19.727   <2e-16 ***
---
Signif. codes:  0 '***' 0.001 '**' 0.01 '*' 0.05 '.' 0.1 ' ' 1

Residual standard error: 7.995 on 537 degrees of freedom
  (229 observations deleted due to missingness)
Multiple R-squared:  0.4202,    Adjusted R-squared:  0.4191
F-statistic: 389.2 on 1 and 537 DF,  p-value: < 2.2e-16
```

First note that, just after displaying the formula, the distribution of the residuals is given. Below, we can see a list of all variables features, not only the value of each coefficient but also the standard error and the significance level. This is not so important yet, with only one variable, but it will be key in the next section. Finally, the *coefficient of determination* R^2[40] which measures the accuracy of the regression line is given as the Multiple R-squared. It measures how much variance of triceps can be explained with the regression line using the variable mass. In this

---

[40]The coefficient of determination R^2 can be obtained as the square of the coefficient of linear correlation r.

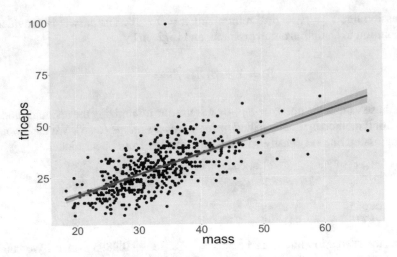

Fig. 5.28 Regression line for `mass` and `triceps`

case, only a 42% is explained so it is not a very explanatory model. The coefficient of determination can be easily called by

```
summary(model)$r.squared
```

```
[1] 0.4201813
```

We can plot the regression with the following code and obtain: Fig. 5.28.

```
ggplot(PimaIndiansDiabetes2, aes(x=mass, y=triceps)) +
  geom_point() +
  geom_smooth(method="lm", se=TRUE)
```

The regression line is the line that best fits to the points in the scatterplot. Given that we define this notion of best to be the one minimizing the residuals, which are the differences between the original values for `triceps` and the ones predicted with the line, it turns out that it is the closest line to the points in vertical distance.

The regression line can be used to make predictions about new values with the `predict()` command. The function is prepared to batch many new observations, so a data frame is expected. However, when passing just one data to be predicted the function looks more complicated, given that a data frame is created for just one value. For example, if we want to predict the value for `triceps` when `mass` is 30, the command is

```
predict(model, data.frame(mass=30), interval="confidence")
```

```
        fit      lwr      upr
1 26.31098 25.57680 27.04515
```

If the argument `interval` is removed, then only the prediction is shown. Other values of that argument are out of the scope of this book.

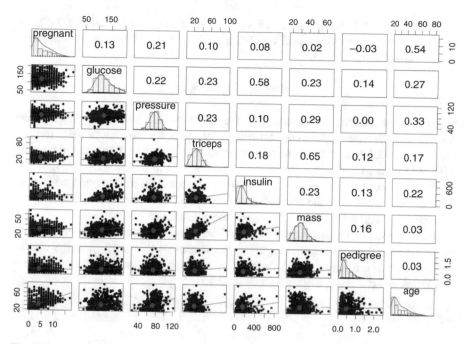

Fig. 5.29 Fancy pairs for Pima Indian variables

Finally, a fancier version of plot(PimaIndiansDiabetes2) can be performed with the help of the library psych and function pairs()

```
library(psych)
pairs.panels(PimaIndiansDiabetes2[, -9],
             method="pearson", hist.col="seashell",
             density=TRUE, lm=TRUE)
```

to produce an enhanced plot of all pairwise linear regressions and histograms of each variable, see Fig. 5.29.

5.3.3 Multivariable Linear Models

In this section, we generalize the regression line to the case of more than one independent variable. That is, given a dataset with variables $Y, X_1, \ldots, X_k$ we want to derive *linear models* for one variable in terms of the others as $Y = \beta_0 + \beta_1 X_1 + \ldots + \beta_k X_k$. This is called the *multiple regression model* or *general linear model*. We fit the model (compute the values of the coefficients) by minimizing the residual sum of squares in a similar way to how β_0 and β_1 were computed for the regression line. However, in this case, the optimization problem comes along with longer calculations that involve matrix calculus, residuals being defined again as the differences between the fitted values $\hat{Y}$ and the actual values of Y. Then, we minimize these residuals and we obtain a matrix formula for the *ordinary least squares* coefficients given by

$$\beta = (X^t X)^{-1} X^t Y ,$$

where X is a matrix whose columns are given by the values of the variables $X_1, X_2, \ldots, X_k$ (and a first column filled with ones for the independent term of the linear model β_0). In the equation, X^t is its transpose and Y is a column vector with the values of variable Y. Some assumptions have to be made for this model to work as, for example, the residuals should follow a normal distribution with mean zero, but they are out of the scope of this book (see [8]). We call the variable Y the *response* and the variables $X_1, X_2, \ldots X_k$ the predictors.

The function to fit linear models by ordinary least squares in **R** goes similarly to the one for the regression line, using the + symbol to specify which variables are used. We fit a model for triceps using mass, glucose, and pressure as predictors and call a summary for this model.

```
model <- lm(triceps ~ mass + glucose + pressure)
summary(model)

Call:
lm(formula = triceps ~ mass + glucose + pressure)

Residuals:
    Min      1Q  Median      3Q     Max
-20.128  -4.878  -0.649   4.540  66.358

Coefficients:
             Estimate Std. Error t value Pr(>|t|)
(Intercept) -6.20853    2.43688  -2.548   0.0111 *
mass         0.95545    0.05410  17.662   <2e-16 ***
glucose      0.02310    0.01171   1.972   0.0491 *
pressure     0.01637    0.03003   0.545   0.5860
---
Signif. codes:  0 '***' 0.001 '**' 0.01 '*' 0.05 '.' 0.1 ' ' 1

Residual standard error: 8.008 on 528 degrees of freedom
  (236 observations deleted due to missingness)
Multiple R-squared:  0.4242,    Adjusted R-squared:  0.4209
F-statistic: 129.7 on 3 and 528 DF,  p-value: < 2.2e-16
```

As before, plenty of information is available. First, the coefficients are listed together with information about the impact of each variable in the response using this model. For example, in our case, we can observe that the variable mass has a coefficient of 0.95545 meaning that, if the rest of the variables do not vary, an increase of one unit in mass implies an increase of 0.95545 in the response variable triceps. This is extended to every variable. Hence, positive coefficients stand for positive impacts and negative coefficients mean that the influence of this variable is negative. Note that the (Intercept) variable (that is, the constant term) has been automatically included. This can be avoided just by including -1 as a variable. Together with the values of the coefficients we have its *standard errors* which are the estimated standard deviations for them, as they come from a sample. The t-value is the statistic used for a student's t hypothesis test in which the null hypothesis is that the corresponding coefficient is zero, and hence *not significant*. The last value in each row is the p-value for this test. The output features significance codes via the use of asterisks, a friendly way of visualizing the p-values to decide at a glance if the relation is rejected at a certain significance level or not. The more $*$ means more significant the variable is, ranging from one to three stars.

The coefficient of determination R^2, which measures how much variance of the response variable Y, is explained by the rest of the variables $X_1, X_2, \ldots, X_k$, is also provided. In this case, `mass`, `glucose`, and `pressure` explain together 42.42% of the total variance of `triceps`. The *adjusted R-squared* is a similar measure for the reliability of the model, but it penalizes the number of predictors. That is, as the number of variables increase, the adjusted R-squared reduces more and more in comparison to R^2. If we include more and more variables, the R^2 increases because we include more information to explain Y although, at some point, the increase is despicable. It is, however, preferable to choose models with a smaller number of predictors (for similar R^2) since they are simpler and more easily interpreted. Adding more predictors should only be done when the reliability of the model increases noticeably. In this sense, the adjusted R^2 helps us to measure if it is worth to include new variables to the model or not. A very small increase in R^2 might turn into a decrease in the adjusted R-squared.

Finally, we observe an F-statistic with its p-value. This corresponds to the *joint significance test*, performed by using a Snedecor's F distribution, where the null hypothesis is that every coefficient is null.[41] That is, $H_0 : \beta_1 = \beta_2 = \ldots = \beta_k = 0$ against the alternative hypothesis that at least one is not equal to zero. In the example the null hypothesis is rejected, since the p-value is very small and hence the coefficients of the model are globally significant.

We often begin with a model using all the available variables of the data frame. To do it, just write the dot symbol as the predictor and **R** understands that all variables (except the one to predict) should be used. We construct a linear regression model for the variable `mass`.

```
model <- lm(mass ~ ., data=PimaIndiansDiabetes2)
summary(model)

Call:
lm(formula = mass ~ ., data = PimaIndiansDiabetes2)

Residuals:
    Min      1Q   Median      3Q      Max
-11.9216  -3.3893  -0.7394   3.0061  21.5470
```

---

[41]Except the first one that corresponds to the independent term.

```
Coefficients:
              Estimate Std. Error t value Pr(>|t|)
(Intercept)  15.112359   1.850813   8.165 4.69e-15 ***
pregnant     -0.228028   0.108476  -2.102   0.0362 *
glucose      -0.004484   0.011429  -0.392   0.6950
pressure      0.103319   0.021873   4.724 3.26e-06 ***
triceps       0.398051   0.025744  15.462  < 2e-16 ***
insulin       0.005744   0.002644   2.172   0.0304 *
pedigree      0.817712   0.760767   1.075   0.2831
age          -0.047715   0.036370  -1.312   0.1903
diabetespos   1.576717   0.659531   2.391   0.0173 *
---
Signif. codes:  0 '***' 0.001 '**' 0.01 '*' 0.05 '.' 0.1 ' ' 1

Residual standard error: 5.01 on 383 degrees of freedom
  (376 observations deleted due to missingness)
Multiple R-squared:  0.5023,    Adjusted R-squared:  0.4919
F-statistic: 48.31 on 8 and 383 DF,  p-value: < 2.2e-16
```

Observe that the variable diabetes is not in the model and instead we find diabetespos. This is so because diabetes was a qualitative variable with pos or neg values and, in order to include it in the model, it has been transformed into a *dummy* variable diabetespos, which is a variable whose values are just 0 and 1 depending on the value of diabetes. In order to check the coding of the new variable we can type the following:

```
contrasts(diabetes)
```

```
pos
neg   0
pos   1
```

Now, we can check that the model is globally significant and that it is able to explain the 50.23% of the variance of mass using all the variables. We can also observe that there are some non-significant variables as glucose, pedigree, and age. We proceed to delete the less significant ones by adding a minus sign in the model call.

```
model <- lm(mass ~ . -glucose - pedigree - age,
            data=PimaIndiansDiabetes2)
summary(model)
```

```
Call:
lm(formula = mass ~ . - glucose - pedigree - age,
                data = PimaIndiansDiabetes2)

Residuals:
     Min       1Q    Median       3Q      Max
-12.1586  -3.3481   -0.6686   2.9879  21.5916

Coefficients:
               Estimate Std. Error t value Pr(>|t|)
(Intercept)  14.498506   1.526983   9.495  < 2e-16 ***
```

```
pregnant      -0.325957    0.082872   -3.933 9.93e-05 ***
pressure       0.095618    0.021376    4.473 1.02e-05 ***
triceps        0.400059    0.025548   15.659  < 2e-16 ***
insulin        0.004908    0.002251    2.180   0.0298 *
diabetespos    1.451142    0.595921    2.435   0.0153 *
---
Signif. codes:   0 '***' 0.001 '**' 0.01 '*' 0.05 '.' 0.1 ' ' 1

Residual standard error: 5.01 on 386 degrees of freedom
  (376 observations deleted due to missingness)
Multiple R-squared:  0.4983,    Adjusted R-squared:  0.4918
F-statistic: 76.69 on 5 and 386 DF,  p-value: < 2.2e-16
```

Observe that the R^2 coefficient has decreased a little bit but the adjusted R^2 coefficient has not. This is because we have removed the non-significant variables, the ones for which the significance tests result in that there is no statistical evidence to claim that they are different from zero.

5.3.4 Non-linear Transformations

It is possible to transform a linear model in order to deal with non-linear relations between the variables and, possibly, improve the model. The possibility that a non-linear relation exists is easily detected by inspecting the pairwise plots of all variables. Recall that linear relations are found when a linear pattern is found in any of the plots. Shapeless clouds of points suggest no relations. A non-linear relation is detected when a curved pattern appears, typically one of the variables growing faster than the other. The PimaIndianDiabetes2 dataset only features the first two kinds of settings, as shown Fig. 5.29. For this section we go back to the diamonds set that we used in Chap. 4. When all relations are plotted, non-linear patterns are found in various plots. We focus on two of them, the relation between x and price and the one between carat and x.

5.3.4.1 Polynomial Transformations

Every time the predicted variable grows faster than the predictor, we might be facing a polynomial relation. In that setting a polynomial regression of order n is fitted. That is, for a pre-specified n we want to find the β_0, β_1, β_2, ..., β_n such that the polynomial

$$response = \beta_0 + \beta_1 predictor + \beta_2 predictor^2 + ... + \beta_n predictor^n$$

is the most reliable fit to the set of observations. Note that, here, we are using the first n powers of the same predictor, and therefore we are using a single variable in this model. This can be performed with lm() specifying in the predictor the polynomial

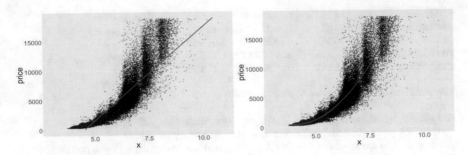

Fig. 5.30 Linear (left) and non-linear (right) regression fitting

that we want to consider, with the command `poly(variable, order)`. For example, we want to predict `price` by using only `x`; this could be done simply with the linear model to obtain rather good results

```
linear.price <- lm(price ~ x, data=diamonds)
summary(linear.price)$adj.r.squared
```

```
[1] 0.7822215
```

but, when representing the scatterplot together with the linear model, it reveals that the underlying nature of the relation has not been fully described. See Fig. 5.30 left that can be generated using this code.

```
ggplot(diamonds, aes(y=price, x=x)) +
  geom_point(alpha=.5) +
  geom_smooth(method="lm") +
  xlim(3, 11) +
  ylim(0, 19000)
```

A polynomial fitting is likely to work here. The shape of the pattern looks like a parabola, so we perform the regression with a second order polynomial using the code

```
poly.price <- lm(price ~ poly(x, 2), data=diamonds)
summary(poly.price)$adj.r.squared
```

```
[1] 0.8591606
```

Not only has the fit improved, if we plot it with

```
ggplot(diamonds, aes(y=price, x=x)) +
  geom_point(alpha=.5) +
  geom_smooth(method="lm", formula=y ~ poly(x, 2)) +
  xlim(3, 11) +
  ylim(0, 19000)
```

the resulting chart in Fig. 5.30 right reveals a clearly better description of the true relationship between `price` and `x`. Observe the usage of the argument `formula` to automatically obtain the correct fitting.

Note that we can combine these polynomial transformations with more variables. Moreover, we can include only certain powers of a variable by setting `I(variable^n)` in the code, which adds just the n-th power of `variable` to the model.

5.3.4.2 Logarithmic Transformations

The other archetypal scenario happens when the predicted variable grows every time slower as the predictor grows. In that case, we are likely facing a logarithmic relation. If we fit a logarithmic model, we find the β_0 and β_1 parameters such that

$$\text{response} = \beta_0 + \beta_1 \log(\text{predictor})$$

is the most reliable fit to the set of observations. Again, this is performed with `lm()` specifying now that the predictor is transformed by using `log()`. For example, we want to predict `x` using `carat` and, as before, a great fit is obtained by the linear model

```
linear.carat <- lm(x ~ carat, data=diamonds)
summary(linear.carat)$adj.r.squared
```

```
[1] 0.9508078
```

Even though the R^2 is extremely close to 1, if we plot Fig. 5.31 left using the code

```
ggplot(diamonds, aes(y=x, x=carat)) +
  geom_point(alpha=.5) +
  geom_smooth(method="lm")
```

we observe that a linear model does not properly describe the relation, especially for high values of `carat`. Since the immense majority of the values are in the area that is correctly described by the linear model, R^2 is really high and a false perception of correct fitting is obtained. However, by fitting a logarithmic regression

```
linear.carat <- lm(x ~ log(carat), data=diamonds)
summary(linear.carat)$adj.r.squared
```

```
[1] 0.9804279
```

the goodness of the fitting is improved a bit and, in addition, if the relation is plotted with the code

```
ggplot(diamonds, aes(y=x, x=carat)) +
  geom_point(alpha=.5) +
  geom_smooth(method="lm", formula=y ~ log(x))
```

we obtain the right-hand side in Fig. 5.31, where we can see how the description of the relation between `carat` and `x` is better explained.

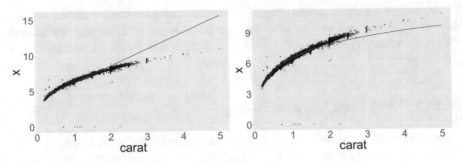

Fig. 5.31 Linear (left) and logarithmic (right) regression fitting

5.3.5 Categorical Variables

We now have a first notion on how to measure the relationship between numerical
variables (correlation), how to obtain a formula relating the variables (regression
line), compute the error generated from fitting (residuals) and make predictions.

However, the models performed so far only predict numerical variables. When a
categorical variable is found as a predictor it is transformed into a dummy variable,
but what if we want to find if there is a relationship between two categorical
variables? Or we aim to fit a multi-variable model with a categorical variable as
a response? As a matter of fact, the Pima dataset was actually designed to study the
`diabetes` using the rest of the variables as predictors. Actually, it is a reference
dataset to evaluate the predictive ability for models, see [19].

We start by comparing two categorical variables by means of the function
`assocstats()` in the `vcd` package. In our example, considering `age` as a
categorical variable, we can count the number of cases of positive and negative
`diabetes` with a contingency table, a table counting the number of cases for each
pair of values.

```
attach(PimaIndiansDiabetes2)
contingency <- table(diabetes, age)
contingency
```

```
age
diabetes 21 22 23 24 25 26 27 28 29 30 31 32 33 34 35 36 37
neg      58 61 31 38 34 25 24 25 16 15 11  7  7 10  5  6 13
pos       5 11  7  8 14  8  8 10 13  6 13  9 10  4  5 10  6
age
diabetes 38 39 40 41 42 43 44 45 46 47 48 49 50 51 52 53 54
neg       6  9  7  9 11  2  3  7  6  2  4  2  3  3  1  1  2
pos      10  3  6 13  7 11  5  8  7  4  1  3  5  5  7  4  4
age
diabetes 55 56 57 58 59 60 61 62 63 64 65 66 67 68 69 70 72
neg       3  1  4  4  1  3  1  2  4  1  3  2  2  1  2  0  1
pos       1  2  1  3  2  2  1  2  0  0  0  2  1  0  0  1  0
age
```

```
diabetes 81
neg   1
pos   0
```

Then we can compute some association measures between them.

```
library(vcd)
assocstats(contingency)
```

```
X^2 df    P(> X^2)
Likelihood Ratio 150.06 51 1.0889e-11
Pearson          140.94 51 2.3070e-10

Phi-Coefficient    : NA
Contingency Coeff.: 0.394
Cramer's V         : 0.428
```

The first two rows show two tests (*Pearson and likelihood ratio*) using the χ^2 distribution. Its corresponding *p*-values are testing the null hypothesis that the variables are independent and, hence, they are not related at all. In this case, we see that we reject the null hypothesis in both tests and therefore there are some relationship between the variables. The following three output rows are coefficients measuring the strength of the association between the variables with several coefficients having values between 0 and 1. The first one is only defined for 2×2 contingency tables and does not make sense in this case. The *contingency coefficient* and the *Cramer's V* measures indicate that no strong association between the variables is present.

A first approach to studying the relation of a categorical variable with quantitative ones is visualizing box plots. As an example, we show the box and whiskers for the two values of diabetes in Fig. 5.32.

```
boxplot(mass ~ diabetes, data=PimaIndiansDiabetes2, ylab="Mass")
```

The chart suggests that there is no relation since both boxplots look similar. In order to actually decide mathematically if there is any association, we perform an *ANOVA* (analysis of variance) test to check if there is statistical significance of

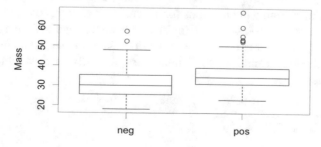

Fig. 5.32 Box plots for mass depending on the value of diabetes

any difference between the variances of the variable mass in the two groups of diabetes and its global variance, with the function aov().

```
aov <- aov(mass ~ diabetes)
summary(aov)
```

```
             Df Sum Sq Mean Sq F value Pr(>F)
diabetes      1   3567    3567    82.4 <2e-16 ***
Residuals   755  32687      43
---
Signif. codes:  0 '***' 0.001 '**' 0.01 '*' 0.05 '.' 0.1 ' ' 1
11 observations deleted due to missingness
```

Hence, we reject the null hypothesis that the variances for each value of diabetes are equal and therefore there is no relationship between the variables. Actually, we can interpret the *p*-value as a measure of correlation between them.

5.3.5.1 Classification

A model where the response variable is categorical is called a *classification model*. The first approach to classification is the *logistic regression* (or *logit*), an extension of the linear model which models the probability of the response variable being 1, instead of modelling the concrete values of the variable.

The binary variable diabetes is coded as 0 for neg and 1 for pos. With this coding, linear regression will be able to give values representing approximately the probability of diabetes taking the value 1. But the model does not understand that it is a probability and, hence, some values could be below 0 or above 1. Then, we use the formula

$$P(Y = 1) = \frac{e^{\beta_0 + \beta_1 X_1 + \beta_2 X_2 + \dots + \beta_k X_k}}{1 + e^{\beta_0 + \beta_1 X_1 + \beta_2 X_2 + \dots + \beta_k X_k}},$$

which does the trick and gives us a value between 0 and 1, which is a probability. The fit of this model is made by a different method called *maximum likelihood* because it has better statistical properties, although it could also be fitted by ordinary least squares, by transforming the previous formula using logarithms into a linear one.

In any case, it is very easy to fit a logistic regression model in **R**. We just need to use the function glm which stands for *generalized linear models* with the argument family=binomial, and we fit a model for the diabetes variable using the other variables.

```
log.model <- glm(diabetes ~ ., data=PimaIndiansDiabetes2,
                 family=binomial)
summary(log.model)
```

```
Call:
glm(formula = diabetes ~ ., family = binomial, data = PimaInd ...
```

```
Deviance Residuals:
    Min        1Q    Median        3Q        Max
-2.7823   -0.6603   -0.3642    0.6409    2.5612

Coefficients:
              Estimate Std. Error z value Pr(>|z|)
(Intercept) -1.004e+01  1.218e+00  -8.246  < 2e-16 ***
pregnant     8.216e-02  5.543e-02   1.482  0.13825
glucose      3.827e-02  5.768e-03   6.635 3.24e-11 ***
pressure    -1.420e-03  1.183e-02  -0.120  0.90446
triceps      1.122e-02  1.708e-02   0.657  0.51128
insulin     -8.253e-04  1.306e-03  -0.632  0.52757
mass         7.054e-02  2.734e-02   2.580  0.00989 **
pedigree     1.141e+00  4.274e-01   2.669  0.00760 **
age          3.395e-02  1.838e-02   1.847  0.06474 .
---
Signif. codes:  0 '***' 0.001 '**' 0.01 '*' 0.05 '.' 0.1 ' ' 1

(Dispersion parameter for binomial family taken to be 1)

    Null deviance: 498.10  on 391  degrees of freedom
Residual deviance: 344.02  on 383  degrees of freedom
  (376 observations deleted due to missingness)
AIC: 362.02

Number of Fisher Scoring iterations: 5
```

As it was the case for linear models, we obtain a summary with the coefficients for the model and the significance of the variables. We observe that there is no coefficient of determination since it does not make sense in this context. To evaluate the predictive ability we need to check at hand if our model would classify the observations of our dataset properly. To obtain the predictions of our model, we use the `predict()` function. If no extra data is supplied, it computes the predictions for the elements of our dataset. Here we need to use the argument `type="response"` to obtain the predicted probabilities of `diabetes` being pos.

```
predictions <- predict(log.model, type="response")
head(predictions)
```

```
         4          5          7          9         14         15
0.02711452 0.89756363 0.03763559 0.85210016 0.79074609 0.71308818
```

We now encode the predictions with the same factors as `diabetes` to check the success of the model. Predictions are labelled with `pos` when the probability is greater than 0.5[42] and we display the *confusion matrix* to test the success and failure of our model.

```
fact.predictions <- rep("neg", length(diabetes))
fact.predictions[predictions > 0.5] <- "pos"
table(fact.predictions, diabetes)
```

```
                 diabetes
fact.predictions neg pos
            neg 378 189
            pos 122  79
```

This matrix allows us to evaluate the capacity of the model of predicting `diabetes` in all the observations. We can compute the overall success by computing the number of correct predictions divided by the total number of observations with

```
mean(fact.predictions == diabetes, na.rm=TRUE)
```

```
[1] 0.5950521
```

Note that we can also compute the success only in the `pos` cases or in the `neg` ones. If we change the threshold to declare a `pos`, both the confusion matrix and the overall success are modified.

```
fact.predictions <- rep("neg", length(diabetes))
fact.predictions[predictions > 0.6]="pos"
table(fact.predictions, diabetes)
```

```
                 diabetes
fact.predictions neg pos
            neg 400 198
            pos 100  70
```

```
mean(fact.predictions == diabetes, na.rm=TRUE)
```

```
[1] 0.6119792
```

Even if the overall performance is better, the success in the `pos` cases is worse. In this case, we probably are more interested in detecting `pos` cases correctly even though some more `neg` cases can be misclassified. It depends on the dataset at hand and the purpose of the researcher to establish the adequate threshold.

Finally, we can use our model to predict new cases of `diabetes` using new values for the rest of the variables. Let us fit a model with less variables to simplify the code.

```
log.model <- glm(diabetes ~ glucose, data=PimaIndiansDiabetes2,
                 family=binomial)
```

---

[42]Although any other threshold can be chosen.

Now, we can answer to the question of whether a woman with `glucose`=138, `mass`=35.5, and `pedigree`=0.325 is going to suffer from `diabetes` with

```
new.data <- data.frame(glucose=138, mass=35.5, pedigree=0.325)
predict(log.model, newdata=new.data, type="response")
```

```
         1
0.473114
```

Our model points out that the probability of such a disease is 0.4818 for that woman.

For a very complete picture of these and much more advanced models we recommend [10, 11].

5.3.6 Exercises

Exercise 5.25 Construct data frames with simulated values for two variables trying to reach a very low linear correlation (for example, in the interval $[-0.05, 0.05]$) and another one for a very high negative one (for example, in the interval $[-1, -0.95]$). Interpret these values. Now, define values for a variable which is very close to a constant value. Try to find values for a second variable making the linear correlation coefficient the closest possible to one. Make a regression line for these two variables and compute the coefficient of determination. Plot and interpret the results.

Exercise 5.26 Consider the `cars` dataset in **R**. It consists on the distances `dist` (in ft) taken by some old cars to stop at some `speed` (in mph). Explore the relationship between the variables. Try to fit the best linear model for `dist` with `speed`. Plot and interpret the resulting model. Hint: Find on the internet the real estimated formula relating these variables; an R^2 of more than 91% can be achieved.

Exercise 5.27 Try to find the best multiple linear model for the blood `pressure` in the Pima Indian dataset using the rest of variables as predictors and interpret it. Compute linear correlations and plot the relations between the variables to discover the best fit.

Exercise 5.28 Perform the same analysis as in Exercise 5.27 to predict the `price` variable of the `diamonds` dataset. Also compute linear correlations and complete your analysis with the corresponding plots.

Exercise 5.29 We can study a numerical variable with a classification model. Consider a new variable for the Pima Indian dataset called `hpressure`, having the values `Yes` for the cases with high blood pressure and `No` for the rest of them, given certain threshold (for example, the third quartile of a medical measure). Fit a logistic regression model to predict `hpressure` and evaluate its achievement in general, and its false positive and negative rates.

5.4   Data Analysis of the `flights` Database

We now perform some data analysis over the dataset `flights` with the tools we have learned in this chapter. We suppose that the database is already preprocessed, as shown in Sect. 3.3.3.

The very first thing to do is to get a global idea of the dataset. We should have a look at the variables we have in this dataset by `head(flights)` and observe that the first three variables are categorical and the rest are numerical. Now that we are aware of the type of variables we have, we begin our study. With a `summary()`, we can obtain an idea of the values that each variable takes.

```
summary(flights)
```

```
        icao24                 callsign                country
Length:3822909        Length:3822909         Length:3822909
Class :character      Class :character       Class :character
Mode  :character      Mode  :character       Mode  :character



  longitude              latitude            altitude.baro
Min.   :-163.42      Min.   :-57.13      Min.   : -960.1
1st Qu.: -89.98      1st Qu.: 31.73      1st Qu.: 3794.8
Median : -12.82      Median : 38.40      Median : 9448.8
Mean   : -22.12      Mean   : 33.98      Mean   : 7668.4
3rd Qu.: 24.10       3rd Qu.: 44.94      3rd Qu.:10972.8
Max.   : 179.07      Max.   : 88.50      Max.   :38648.6
 velocity               track               vertical.rate
Min.   :   0.0       Min.   :-172.63     Min.   :-1541.140
1st Qu.: 151.2       1st Qu.:  92.07     1st Qu.:   -0.650
Median : 206.9       Median : 191.10     Median :    0.000
Mean   : 189.8       Mean   : 183.47     Mean   :   -0.193
3rd Qu.: 236.3       3rd Qu.: 270.83     3rd Qu.:    0.330
Max.   :2384.5       Max.   : 359.90     Max.   : 7764.780
 time
Min.   :     0
1st Qu.:203400
Median :395100
Mean   :391961
3rd Qu.:585890
Max.   :777600
```

In order to understand the relation among the numerical variables we can use the `corrplot()` function[43]

```
library(corrplot)
corrplot(cor(flights[, -c(1, 2, 3)],
          use="pairwise.complete.obs"))
```

---

[43]Recall that we can compute the correlation coefficient between every pair by using the function `cor()`, but sometimes a plot is more informative, direct, and useful.

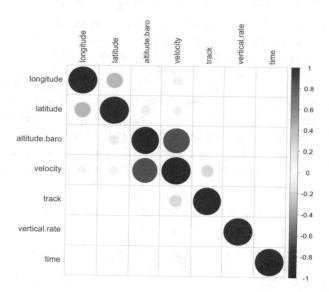

Fig. 5.33 Linear correlations of the numerical variables of the `flights` dataset

and obtain in a very graphic way the linear correlations between them, as shown in Fig. 5.33.

We can appreciate that the unique strong correlation is the one between the variables `velocity` and `altitude.baro` which is positive. Other positive but weaker relations can be observed between the variable `velocity` and the variables `longitude` and `latitude`. Also, there is a positive relation between `altitude.baro` and `latitude`. We find negative correlations between the pair `longitude` and `latitude` and between `velocity` and `track`. The correlations between every pair of variables are useful to detect which variables can be more suitable to be used when constructing linear models.

We now study the association between two categorical variables, `callsign` and `country`, by means of the `vcd` library.

```
library(vcd)
assocstats(table(flights$callsign, flights$country))


                      X^2        df P(> X^2)
Likelihood Ratio  20977515 21686635        1
Pearson          532547665 21686635        0

Phi-Coefficient    : NA
Contingency Coeff.: 0.996
Cramer's V         : 0.98
```

The association is so good that it might even be surprising at first. On a second thought, though, callsigns are always the same for the same flights. And since the same flight always departs from the same country, the association should be perfect. The fact that both the contingency coefficient or Cramer's V are not exactly one is explainable due to errors from the transmission that are transcribed to the dataset.

Next we implement some linear models. We first try to see if the velocity of a flight can be determined using altitude.baro which should be somehow evident. We already know that those variables have a positive linear correlation. To fit the linear model (a regression line in this case, with only one predictor) we use the lm() function and extract the summary() of the fitted model.

```
model.velocity <- lm(velocity ~ altitude.baro, data=flights)
summary(model.velocity)


Call:
lm(formula = velocity ~ altitude.baro, data = flights)

Residuals:
    Min      1Q  Median      3Q     Max
-565.89  -23.36   -0.06   24.00 2150.63

Coefficients:
                Estimate Std. Error t value Pr(>|t|)
(Intercept)    8.749e+01  3.967e-02    2205   <2e-16 ***
altitude.baro  1.334e-02  4.564e-06    2922   <2e-16 ***
---
Signif. codes:  0 '***' 0.001 '**' 0.01 '*' 0.05 '.' 0.1 ' ' 1

Residual standard error: 36.53 on 3822907 degrees of freedom
Multiple R-squared:  0.6908,    Adjusted R-squared:  0.6908
F-statistic: 8.54e+06 on 1 and 3822907 DF,  p-value: < 2.2e-16
```

We obtain a good model, with 68% of thee variability of velocity determined by altitude.baro. More variables can be introduced to check whether the model can be improved or not. It looks reasonable to use latitude and longitude as predictors since the position on the globe where the plane is located might be a good reason to affect the speed and they have some correlation with the response variable velocity. Nevertheless, this model does not improve the previous model more than a 1% and thus the inclusion of those variables is not justified.

```
model.velocity <- lm(velocity ~ longitude + latitude +
                          altitude.baro,
                          data=flights)
summary(model.velocity)


Call:
lm(formula = velocity ~ longitude + latitude + altitude.baro,
data = flights)

Residuals:
    Min      1Q  Median      3Q     Max
-567.15  -23.60   -0.88   23.77 2146.62
```

```
Coefficients:
                Estimate Std. Error t value Pr(>|t|)
(Intercept)     8.695e+01  4.805e-02 1809.75   <2e-16 ***
longitude       5.542e-02  2.456e-04  225.60   <2e-16 ***
latitude        6.625e-02  9.723e-04   68.14   <2e-16 ***
altitude.baro 1.327e-02  4.573e-06 2902.15   <2e-16 ***
---
Signif. codes:  0 '***' 0.001 '**' 0.01 '*' 0.05 '.' 0.1 ' ' 1

Residual standard error: 36.29 on 3822905 degrees of freedom
Multiple R-squared:  0.6948,    Adjusted R-squared:  0.6948
F-statistic: 2.901e+06 on 3 and 3822905 DF,  p-value: < 2.2e-16
```

We can try a new model by using instead the variable track since a negative relation was previously found.

```
model.velocity <- lm(velocity ~ altitude.baro + track,
                     data=flights)
summary(model.velocity)
```

```
Call:
lm(formula = velocity ~ altitude.baro + track, data = flights)

Residuals:
    Min      1Q Median     3Q     Max
-584.87  -20.64    0.89  22.42 2146.45

Coefficients:
                Estimate Std. Error t value Pr(>|t|)
(Intercept)     1.082e+02  4.977e-02  2173.2   <2e-16 ***
altitude.baro 1.329e-02  4.340e-06  3063.4   <2e-16 ***
track          -1.110e-01  1.743e-04  -636.9   <2e-16 ***
---
Signif. codes:  0 '***' 0.001 '**' 0.01 '*' 0.05 '.' 0.1 ' ' 1

Residual standard error: 34.73 on 3822906 degrees of freedom
Multiple R-squared:  0.7204,    Adjusted R-squared:  0.7204
F-statistic: 4.926e+06 on 2 and 3822906 DF,  p-value: < 2.2e-16
```

With the addition of this variable, a 3.35% more of the variability of the velocity is explained. Recall that the simpler the model the easier to use and interpret. So, in each case, the researcher should decide if it is preferable to add more variables to the model or not, trying to find a trade-off between simplicity and accuracy.

Finally it would be reasonable to think that the vertical.rate could be predicted by means of latitude, longitude, altitude.baro, and velocity since knowing the position and speed might be enough to determine whether the plain is going up or down and how much. However, no relation is found whatsoever.

```
model.vertical <- lm(vertical.rate ~ longitude + latitude +
                     altitude.baro + velocity, data=flights)
summary(model.vertical)
```

```
Call:
lm(formula = vertical.rate ~ longitude + latitude +
altitude.baro + velocity, data = flights)

Residuals:
   Min       1Q  Median       3Q      Max
-1541.0    -0.6     0.0      0.4   7764.9

Coefficients:
                Estimate Std. Error t value Pr(>|t|)
(Intercept)   -1.134e+00  2.005e-02 -56.544  < 2e-16 ***
longitude     -3.130e-04  7.573e-05  -4.133 3.59e-05 ***
latitude      -1.871e-03  2.980e-04  -6.280 3.40e-10 ***
altitude.baro -2.266e-06  2.507e-06  -0.904   0.366
velocity       5.347e-03  1.566e-04  34.134  < 2e-16 ***
---
Signif. codes:  0 '***' 0.001 '**' 0.01 '*' 0.05 '.' 0.1 ' ' 1

Residual standard error: 11.11 on 3822904 degrees of freedom
Multiple R-squared:  0.0009354,  Adjusted R-squared:  0.0009343
F-statistic: 894.8 on 4 and 3822904 DF,  p-value: < 2.2e-16
```

The reason for this could be that the vertical rate is a variable that changes quickly (contrarily to velocity that changes smoothly) and the exact position of the plane, no matter where on Earth is located, has a high dose of randomness.

This dataset allows for further and deeper analysis using *clustering*, *Principal Components Analysis*, and many other more advanced multivariate techniques that go beyond the scope of this book.

References

1. Ash, R.B. *Real Analysis and Probability Academic Press*. Academic Press, Massachusetts, USA, 1972.
2. Chihara, L. and Hesterberg, T. *Mathematical statistics with resampling and R*. Wiley Online Library, New Jersey, USA, 2011.
3. De Finetti, B. *Theory of probability: A critical introductory treatment*. John Wiley & Sons, New York, USA, 2017.
4. Fayyad, U.M. and Irani, K.B. The attribute selection problem in decision tree generation. In *Proceedings of the tenth national conference on Artificial intelligence*, pages 104–110. AAAI Press, 1992.
5. Fischer, H. *A history of the central limit theorem: From Classical to Modern Probability Theory*. Springer Science & Business Media, Berlin, Germany, 2010.
6. Fisher, R.A. XV. - The correlation between relatives on the supposition of Mendelian inheritance. *Earth and Environmental Science Transactions of the Royal Society of Edinburgh*, 52(2):399–433, 1919.
7. Gini, C. *Variabilità e mutabilità. Contributi allo studio dele relazioni e delle distribuzioni statistiche*. Studi Economico-Giuridici della Università di Cagliari, Torino, Italy, 1912.
8. Greene, W.H. *Econometric analysis*. Prentice Hall, New Jersey, USA, 2000.
9. Hald, A. *A history of parametric statistical inference from Bernoulli to Fisher, 1713-1935*. Springer Science & Business Media, Berlin, Germany, 2008.

10. Hastie, T., Tibshirani, R. and Friedman, J.H. *The elements of statistical learning: data mining, inference, and prediction.* Springer, Berlin, Germany, 2009.
11. James, G., Witten, D., Hastie, T. and Tibshirani, R. *An Introduction to Statistical Learning: With Applications in R.* Springer Publishing Company, Incorporated, New York, USA, 2014.
12. Jeffreys, H. *The theory of probability.* Oxford University Press, Oxford, UK, 1998.
13. Johnson, M.E. and Lowe, V.W. Bounds on the sample skewness and kurtosis. *Technometrics,* 21(3):377–378, 1979.
14. Katsnelson, J. and Kotz, S. On the upper limits of some measures of variability. *Archiv für Meteorologie, Geophysik und Bioklimatologie, Serie B,* 8(1):103–107, 1957.
15. Kenney, J.F. and Keeping, E.S. *Mathematics of Statistics.* D. Van Nostrand Company Inc, New Jersey, USA, 1951.
16. Kolmogorov, A.N. and Bharucha-Reid, A.T. *Foundations of the theory of probability: Second English Edition.* Courier Dover Publications, New York, USA, 2018.
17. Ramsey, F.P. Truth and probability. In *Readings in Formal Epistemology,* pages 21–45. Springer, Berlin, Germany, 2016.
18. Savage, L.J. *The foundations of statistics.* Courier Corporation, New York, USA, 1972.
19. Smith, J.W., Everhart, J.E., Dickson, W.C., Knowler, W.C. and Johannes, R.S. Using the adap learning algorithm to forecast the onset of diabetes mellitus. In *Proceedings of the Annual Symposium on Computer Application in Medical Care,* page 261, Maryland, USA, 1988. American Medical Informatics Association.
20. Stigler, S.M. *The history of statistics: The measurement of uncertainty before 1900.* Harvard University Press, Massachusetts, USA, 1986.
21. Venn, J. *The logic of chance.* Courier Corporation, New York, USA, 2006.
22. Von Hippel, P.T. Mean, median, and skew: Correcting a textbook rule. *Journal of Statistics Education,* 13(2), 2005.
23. Von Mises, R. *Probability, statistics, and truth.* Courier Corporation, New York, USA, 1981.
24. Westfall, P.H. Kurtosis as peakedness, 1905–2014. RIP. *The American Statistician,* 68(3):191–195, 2014.

R Packages and Functions

Printed in the United States
By Bookmasters